Forest H. Belt's

Easi-Guide to CITIZENS BAND RADIO

HOWARD W. SAMS & CO., INC.
THE BOBBS-MERRILL CO., INC.
INDIANAPOLIS · KANSAS CITY · NEW YORK

FIRST EDITION

FIFTH PRINTING—1975

Copyright © 1973 by Howard W. Sams & Co., Inc., Indianapolis, Indiana 46268. Printed in the United States of America.

All rights reserved. Reproduction or use, without express permission, of editorial or pictorial content, in any manner, is prohibited. No patent liability is assumed with respect to the use of the information contained herein. While every precaution has been taken in the preparation of this book, the publisher assumes no responsibility for errors or omissions. Neither is any liability assumed for damages resulting from the use of the information contained herein.

International Standard Book Number: 0-672-20960-8
Library of Congress Catalog Card Number: 73-75081

Preface

Something about mobile two-way radio fascinates people. It seems little short of magic to ride along in your car and talk with someone miles away. Squeezing the button of a communications microphone brings an unusual thrill of power.

Certain "lucky" people use mobile radio regularly. It's part of their work. Police, for example, never go on duty without two-way radio. Taxicabs have two-way radios. Airplane pilots use them. Owners of yachts and small boats can have two-way radio. All these people can talk to others from a distance without hunting a telephone. But the radio equipment they use to do it is expensive.

Two-way radio holds such fascination for some people that they make a hobby of it. They are called radio amateurs or hams. Many years ago, during the infancy of electronic communications, the governments of several countries set aside bands of radio frequencies expressly for hobbyists and experimenters. In the United States, this is officially labeled the Amateur Radio Service. A ham is authorized to talk by radio only with other hams. He can hold conversations with a ham down the street or on the other side of the world. A ham can communicate by radio from his automobile or home. He can use voice or code. Of all the two-way radio users, hams probably derive the most fun from their communications.

But regulations always limited two-way radio. Either you had to be employed in some service authorized to use it or else learn enough technically to qualify you for a ham license. Then, in 1958, came class-D Citizens Radio. The frequencies that were allocated to this service are practical for communication between automobile and home. Finally, the average person could have personal two-way radio service from his car to his home or to another car. And the equipment was not too expensive.

You can have Citizens Band radio if you're a citizen of the United States and at least 18 years of age. You merely file a formal application and pay a license fee. You can operate as many CB radios as you want. Your wife can have a unit in her car, and so can each of your youngsters. If you operate a farm or other business, you can have a unit in the office, at the barn, on the tractor—anywhere you need two-way communications.

There is considerable satisfaction in using CB radio. It was never and is not a hobby service. Illustrations throughout this book show how you can use CB radio in your business or for private use. You see typical equipment and how to use it properly. Legal rules are explained. You will find in these pages how to get your license. Photos show how to install CB equipment at home or in your automobile. There is even a chapter on how to maintain equipment at peak operating condition.

Numerous people and organizations helped make this book thorough and meaningful for you. I especially thank Larry Belt, Jerry Blythe, Bob Cornelius, Linda Cummings, Stuart Haag of Business Radio Inc., Jay Hendrix of J'town Radio Shack, Don Lange, Jim Newman, Rodney Renfrey, John Roppel and his family, Charles Smith of Smith Two-Way Communications, J. B. Wathen III of Mobile Communications Inc., the Consulate General of Chile in New York, Kentucky Rescue Association, REACT, Avanti R & D, Cush-Craft, Dynascan Corp. (Cobra), Electro-Voice, Heath Company, Linear Systems Inc. (SBE), Mosley Electronics, Motorola, Pathcom (Pace), Pearce-Simpson, Radio Shack, Regency, Shakespeare, and Turner Co.

FOREST H. BELT

Contents

CHAPTER 1

Citizens Radio Service 7
What class-D radio is—Volume VI, Part 95 of FCC Rules and Regulations—Who can use your CB units and for what—Channel 9—Rules that limit CB range—Proposed class-E Citizens Radio

CHAPTER 2

Speaking the Language 25
CB words you should know: Transceiver—Kinds of walkie-talkies—Linear amp—Watts—Input and output power—Sensitivity—Skip—Frequency—Megahertz—Synthesizer—Selectivity—AM and SSB—Modulation limiter—Noise limiter—Squelch

CHAPTER 3

Kinds of CB Equipment 39
Base stations—Single-sideband and a-m models—Six-channel vs 23-channel—Tube vs transistors—Mobile units—Built-in pa system—All voltage CB rigs—CB and Part 15 walkie-talkies—Microphones—A CB transceiver kit

CHAPTER 4

Antennas That "Reach Out" 53
Groundplane, dipole, and gamma-match verticals—Beams—Five-eighths-wave verticals—A coaxial CB antenna—Unusual designs—Coaxial lead-in cable and its impedance—Mobile antennas and mounts—Loading—A co-phased dual antenna for auto use—Connectors

CHAPTER 5

Shopping for CB Gear 67
Mail order companies—Chain outlets—How to shop various kinds of stores—Communications specialists—Electronic parts distributors

CHAPTER 6

Getting Your CB License 75
FCC Form 505 instructions—Ordering required Volume VI—How to fill in the application form—What to do with completed application—Where to get forms—The license and when to renew

CHAPTER 7

Installing Your CB Base Station 87
Appearances—Components of a base station hookup—Electric power—Installing an antenna—Lightning protection—Running the coaxial lead-in cable

CHAPTER 8

Installing the Mobile Rig 105
Mounting the transceiver—Connecting the power source—A trunk-rim antenna mount—Attaching and routing coaxial cable—Installing a solderless connector—The final checkout—Transmitter Identification Form 452-C

CHAPTER 9

Operating on the Citizens Band 125
Usefulness of CB—Illegal operation—Mike technique—How to call and answer CB stations—The APCO 10-code—Priority of communications—Public-service CB groups—REACT and Kentucky Rescue Association—Your responsibility for transmitter emissions

CHAPTER 10

Adjustments and Maintenance 139
Precautions—Keeping your equipment working properly—Transmitter faults—Licensed technicians—Adequate servicing facilities

Chapter 1

Citizens Radio Service

Personal communications for any U.S. citizen 18 years of age or older—that's the purpose of Citizens Band (CB) radio. You do not have to be engaged in any particular occupation nor demonstrate any unusual need for radio communications. You don't have to know much more about electronics than how to squeeze a mike button. CB radio is yours if you want it, are willing to make a formal application, and pay the license fee.

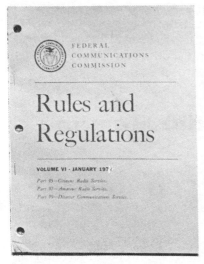 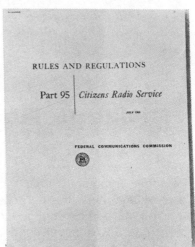

As with almost anything in organized society, certain laws must be followed. Rules and Regulations are promulgated by the Federal Communications Commission (FCC), which oversees all communications in this country. The R & R are your bible for using the Citizens Radio Service.

The FCC was established by the Communications Act of 1934. Essentially, whatever you can or cannot do with communications is because the FCC says yes or no. Wisdom therefore dictates that you familiarize yourself with the rules applicable to the Citizens Radio Service if you plan to use it.

All of the rules, regulations, and instructions for the legal use of CB Radio are included in Volume VI of the FCC Rules and Regulations. Some of what's in Volume VI relates to other communications services. You need only learn about the Citizens Radio Service. That is in Part 95, the volume shown on the right. Just send a check or money order for $5.35 to Superintendent of Documents, Government Printing Office, Washington, DC 20402.

When you apply for a license, one of the stipulations is that you have a copy of and have read Part 95 of the FCC Rules and Regulations. This is so the FCC knows that you understand what is expected of you and what is taboo. That way, you have no excuse to misuse your radio.

You can operate more than one CB transmitter. That is, you can have one at home, one in your car, one in your wife's car, and several in your business vehicles. Even for one, however, you have to apply for a CB station license. You do that on FCC Form 505. You have to fill out the application blank in detail (Chapter 6 tells exactly how). Then mail it, along with a $4 (at this printing) license fee, to the Federal Communications Commission office in Gettysburg, PA 17325.

You need only one license for all your transmitters. However, you must say on your Form 505, how many you want to operate. Ordinarily, you apply for the number you expect to use, plus one or two for good measure. That way you can add that extra unit at the office, or in another car, without having to file a revision to your license (another $8).

The license you get is merely a large card that gives your address and stipulates how many transmitters are authorized under your license. It also gives you a specific station call sign which must be used to identify communications by your station. A CB call sign consists of three letters followed by four numbers.

Your license lasts five years. The expiration date is listed on it. You apply for renewal sometime during the last month or so of its validity. Renewal costs another $4 and is good for another five years. You again use Form 505.

The license is for you. It does not cover specific equipment. You cannot transfer your license to someone else in case you decide to sell all your CB equipment. Nor can it be transferred to some other member of your family—although as long as they live at your house they can use your call sign for personal communications with each other or with your base station at home or office.

Who can use your Citizens Band radio legally? First of all, of course, you can. So can anyone who normally uses your radio-equipped vehicle, provided you have given permission.

Members of your family can use CB for any personal communications within the limits set forth in Rules and Regulations, Part 95. Your wife can use it when she goes to the store. Your son can use it when he goes to school. Either can use it when they're together somewhere in the car. In multiple-car families, it's common to have a CB unit in every car.

Hence, if a family member runs into some kind of trouble away from the house, he has only to pick up the microphone and call for help. As a matter of fact, using certain channels, he can call to any other CB station or mobile unit for help.

Citizens Band radio is a downright convenient addition to the safety equipment on any car. You'll see why as you proceed through this book.

Someone who works for you can use Citizens Band radio under your license, provided it is in a vehicle used in connection with business or *your* personal affairs. For example, the truck an employee uses for deliveries or to pick up supplies can have a CB radio installed. The vehicle could be a truck, car, motorcycle, or bicycle. Actually, the guy could carry a CB walkie-talkie.

However, if your employee wants CB in his own car, he has to get his own license. A salesman who works for you could have a CB unit of yours in his personal car. But he couldn't legally use it for his own personal communications—as for example, to *his* home. For that he needs a license of his own.

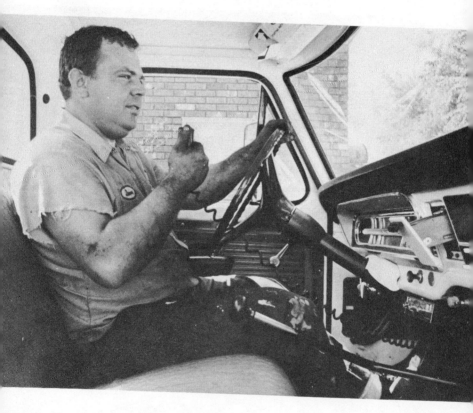

Here, as an example, is the lady of the house confirming the time supper ought to be ready. This kind of communication constitutes what the Citizens Band expressly was activated for. Imagine the convenience. Time saved is only one factor. The sheer relief of knowing why he's late can be worth a lot.

With the Citizens Band, radio communications becomes a family affair. Out on the road or in distant parts of town, this CB family can nevertheless keep in touch. Members can plan where to meet or time their errands to arrive home about the same time.

Perhaps the son needs a few more instructions regarding some item he has to pick up at the hardware store. Or, when he's expected home early, he might relay the news of a traffic jam or being caught on the wrong side of a slow-moving train.

For whatever the purpose, car-to-car and car-to-home communications among a family is one of the more fascinating and useful facets of CB radio.

What begins as a joyful afternoon doesn't always end that way. Safety in times of emergency is only one of the more useful facets of Citizens Band radio. Stranded on the highway, this young lady uses her CB unit to summon aid from home.

There's one CB channel (out of 23) set aside explicitly for emergency communications like this. That's channel 9. This channel is to be used for no other purpose than emergency communications from automobiles. It can be used for guidance—that is, directions on how you can find your way somewhere or what to do about some specific emergency. But channel 9 is not for ordinary conversation.

Again, Citizens Band, used properly, offers more than mere convenience.

Communication by CB radio is not confined now to family or business employees, as it was at first. Provided you adhere to certain limitations, you are free to talk by radio with friends and acquaintances who have their own CB licenses. Here are some of the limitations:
1. Communications between stations not under the same license are restricted to channels 10 through 15 and channel 23. That keeps channels 1 through 8 and 16 through 22 clear for regular CB communications. CBers who interfere by chatting with each other on regular channels operate illegally.
2. Station-to-station conversations can't last longer than 5 minutes. And then the operators must wait at least 5 minutes before resuming conversation. This ruling gives others a chance to use the seven station-to-station channels.
3. Idle chit-chat is forbidden on the CB channels. Those who do it operate illegally. FCC rules limit your talk to necessary communications.

Your most likely application of channel 9 will occur when your car breaks down. Engines overheat on the highway, tires go bad, transmissions fail, and batteries give out on cold winter mornings after a stall in traffic. All these offer legitimate excuse to transmit a request for help on channel 9. Service station owners often equip truck and office with Citizens Band radio and do their major listening on channel 9—the "stranded-motorist" channel.

You can inquire for directions too. Suppose you're in a strange town and can't seem to locate a street. A call on channel 9, assuming it's heard, can bring you instructions from a native on how to proceed. That CB operator can literally "talk you in" to where you're headed. If you're thoughtful of others who might need assistance, you'll start your emergency conversation on channel 9 but then switch to another channel that's not busy.

You can transmit requests for help, or reports of an accident, on any channel. But confine station-to-station conversation, even for guidance, to channels 10 through 15.

Many clubs and organizations encourage members to install CB radio in their automobiles. Then, when they put together picnics, jamborees, and other outings, club members can maintain contact.

The sign in this photo illustrates how the National Campers and Hikers Association helped members find their way to one gigantic summer gathering. NCHA groups from all over the U.S. converged on a spot in central Indiana. Direction signs like this were posted on all major highways leading toward the encampment. Most important, the signs indicated a CB channel to use as members neared their destination. The campsite was remote, and some drivers had to be directed the last few miles.

The NCHA group chose channel 11 more or less arbitrarily. They had no exclusive right to its use. Yet, every CB-equipped member knew where to listen for guidance. This is only one example.

Horse shows, flying clubs, sports-car rallies, and dozens of other activities offer legitimate opportunities for CB to serve a major function.

The preceding pages paint a versatile picture of how Citizens Band radio can make your life easier and more fun. A few restrictions are expressed. You should familiarize yourself with *all* the rules. Procure a copy of Part 95 and read it. The FCC sends a small booklet "How to Use CB Radio" with your license. Study that; its format is programed and easy to comprehend. Illustrations clarify important regulations.

The next few pages bring you some insight into other key regulations . . . some you may prefer to know about *before* you go out and buy Citizens Radio equipment.

For example, the rule about power. Your CB radio cannot legally transmit more than 4 watts of radio frequency (rf) power. Where the regulations set a 5-watt limit, they refer to *input dc power* in the final amplifier stage of the transmitter. Allowing for efficiency losses, that translates to no more than 4 watts of *rf output power*. Inefficient CB transmitters put out less than 4 watts rf power.

The object is to limit distance, so more people can enjoy the use of Citizens Band radio.

There's more to this distance-limiting approach. You may know that the height of your CB antenna has a direct influence on how far CB radio reaches. Radio waves like those for CB travel line-of-sight. They work their way around a few corners (as of tall buildings) and through some trees and over a few hills. But all those obstructions affect the distance you can talk to or hear other CB stations. The higher you put the antenna, the further your CB radio waves travel before being soaked up or blocked out.

But FCC rules include a 60-foot height limitation. Mounted on your house, the tallest tip of your antenna must extend no more than 20 feet above the tallest projection of the building (the chimney, for example). Mounted on the ground, your antenna can reach upward as far as 60 feet—including the mounting pole *and* the antenna.

Besides limiting distance, this rule prevents well-to-do CB owners from erecting towers that could constitute a hazard to aircraft. The FCC and FAA haven't time to check each license application for conformance with Civil Air Regulations.

These limits on distance affect communications to your own vehicles. CB is intended for short-range use, but in rugged terrain the 4-watt and 20-foot rules cut effective distance to an impractical minimum.

You have some recourse. Manufacturers of CB antennas have designed and now sell models that build up effective radiated rf power. Nothing in the FCC regulations prevents this, and it's one way to push your CB transmissions (and reception) a little farther.

The "gain" type of CB antenna usually is a *beam*—a lineup of vertical rods spaced along a round metal boom that points in the direction you want to communicate best. Or, the design may be a *quad*—like the one pictured here. It may be some variety of specially designed vertical whip—such as a ⅝-wavelength design—which extends CB range farther than the simple *groundplane* antenna on page 20.

A "gain" antenna can double or triple the effective radiated power (erp) from your CB station. It also reciprocally improves your reception over the same distance. However, the 60-foot rule reduces to 20 feet for beam antennas. You may not gain all you hope to.

The ultimate distance limitation for Citizens Band radio is 150 miles. That's much further than ordinary CB can reach. Usual distance, with an efficient transmitter feeding a "gain" antenna on a tall house, is 10 to 20 miles. In flat country, or from hilltop to hilltop, you might reach 30 miles.

The 150-mile rule contends with a phenomenon called *skip*. Radio waves of the frequency used for CB (around 27 MHz) can bounce off the ionosphere and return to earth several hundred miles away from the source. This makes possible communications with far-distant stations on days when skip is "good." Such communications are *strictly forbidden* by FCC regulation (see paragraph 83b of Part 95).

The temptation may be great. When you're sitting in Florida and hear a station in Alaska, the sheer challenge of such uncommon distance (this is at least double- or triple-skip) attracts CB operators. Don't give in. The FCC may be listening, and you stand the chance of a citation, fine, or even jail sentence—and of having your license revoked.

As this book goes to press, a new kind of Citizens Radio Service is in the planning stage. It's called *class E,* to distinguish it from the existing 23 class-D channels.

The fate of class-E Citizens Band radio rests with the Federal Communications Commission. The proposal specifies 80 channels at much higher frequencies than are used presently. Class-D channels, as you now know, operate near 27 MHz; class E would operate around 220 MHz.

That choice of frequency has two attractions: (1) no skip interference and (2) room for more channels. The stated objective is to provide channels that are not so crowded, for the serious user of Citizens Radio Service.

Class-E CB two-way radio equipment would differ from the class-D units you see throughout this book. Of course, no class-E models even exist yet, although they could by the time you're reading this.

However, they will look something like the commercial two-way radio pictured on this page. The cost will be higher than for class-D gear, although manufacturers figure they can bring the cost down to around $150–$250 per unit if enough serious CBers move to class-E operation. In the initial stages, costs may run higher than $300 per unit. Likely, only business operators of CB will use it for some time—but that's precisely what the proponents of class E count on.

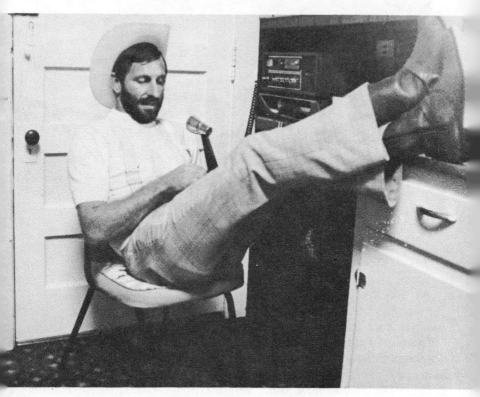

One last word about rules. Citizens Band radio was never intended as a hobby, although most users treat it that way. The Federal Communications Commission ("the big double-C," some CBers irreverently call it) regularly sends monitoring crews out to catch offenders. Although obscene language draws the toughest clamp-down from FCC enforcement teams, overpowered stations, lapsed licenses, and idle chatter capture plenty of their attention.

Pitfalls in CB operation abound, particularly if you listen to the run-of-the-mill CB operator. Some get away with flagrant violations. Other just can't resist using the band for "ragchew" sessions, violating both nature-of-communications and time-limit rules—and often channel-use allocations.

But, if you approach the Citizens Radio Service with serious intent and use it for its planned purpose, you'll avoid trouble and enjoy your own personal communications system. The remaining chapters of this book show you how to do both.

Chapter 2

Speaking the Language

Sitting over your CB rig with a dictionary in hand won't teach you the slang bandied about among CBers. But you should know some of the words you'll hear. Several of them are technical. An ordinary dictionary might not have them anyway.

Some of the words peculiar to Citizens Radio Service already have been used and explained briefly in Chapter 1. Others are scattered throughout the remaining chapters. Here are a few you won't find elsewhere or that deserve extra emphasis or explanation.

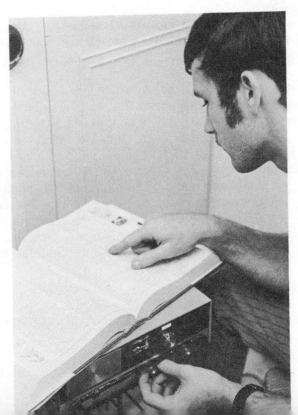

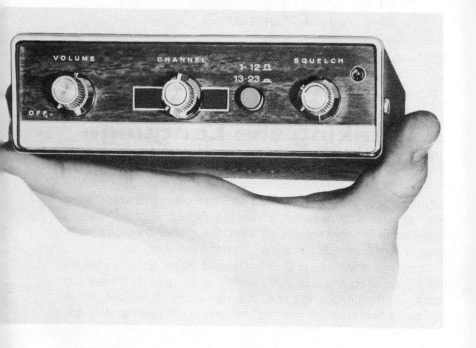

Maybe you never heard the word *transceiver* before. It applies only to two-way radio. The word comes from combining *transmitter* and *receiver*. But there's a small subtlety in the term.

In commercial two-way radios, which are costly, the transmitter and receiver are separate. That is, they're built into the same box but are on independent chassis. Each operates completely on its own. These two-way units are called transmitter/receivers.

For CB, and for some other communications services, certain circuits and stages are built common to both transmitting (you do the talking) and receiving (you hear someone else talking). This saves manufacturing expense and reduces what you have to pay for a unit. This share-the-innards design goes by the name of *transceiver*. All models presently built for Citizens Radio are of the transceiver design.

Handheld transceivers go by the popular name of *walkie-talkie*. A tiny transmitter and receiver (again, with some circuits common) fit into a package that can be easily carried. Batteries furnish operating voltage.

Two kinds of handheld transceivers flourish. One *is not* a true CB transceiver. That's the cheap "toy" walkie-talkie that kids buy or get for Christmas. It is authorized under a special low-power radio service. Although this low-cost walkie-talkie often uses frequencies assigned to Citizens Band radios, you cannot legally communicate with one through your CB rig.

The true CB walkie-talkie sells for more because Citizens Radio Service rules exert stringent equipment requirements. The handheld CB transceiver pictured here includes rechargeable battery, and works on six different channels. It also puts out much more rf power than a "Part 15" walkie-talkie.

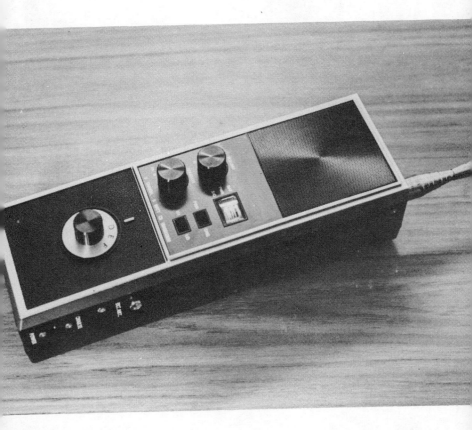

Here's a piece of equipment you'll hear a lot about as you listen to some users of CB radio. The most important fact to remember about it is: IT IS ILLEGAL for CB radio use.

It's a *linear amplifier,* and CBers call it "linear" for short. A linear boosts rf power to a level far above the legal limit for Citizens Band radio. Whereas a CB rig should feed no more than 4 watts to its antenna, the linear amplifier may boost that to 25, 40, or even 60 watts.

Ironically, that doesn't really push the CB signal much farther, because CB radio waves travel mainly line-of-sight. But a high-powered signal blankets the area it does cover. Unfairly, it covers up efforts at communication by legal CB operators. And, for the illegal "skip" operator, the stronger signal bounces a little better than a legal signal.

Linear amplifiers have legitimate uses in other radio services, such as Amateur Radio or Business Radio. But they are illegal to sell or use for Citizens Band stations.

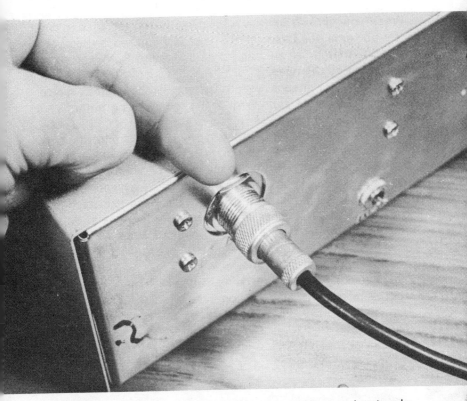

You may wonder what this *watt* is that you can send out only four of. The watt is a measure of power—in this instance, of radio-wave (called rf for *radio frequency*) power.

A licensed technician measures the output of your transmitter in watts of rf power. If your transmitter puts more than 4 watts out through that antenna connector, the technician must readjust the output circuits to keep your transmitter legal. He's setting *output* power.

Another wattage term, *input* power, appears in FCC regulations. It refers to a different way of measuring transmitter power, but again the unit is the watt. Input power refers to the amount of voltage and current used by the final transistor or tube stage in your transmitter. This is the stage just before the output connector. Power measured by this indirect method cannot exceed 5 watts. How much rf wattage comes from the output stage depends on its efficiency. If input power is 5 watts and output rf power is 4 watts, efficiency is 0.8 or 80 percent.

When your mobile units are far away in the distance, transmitter power is one factor in how well you can reach them, and they, you. Another is your base-station antenna. Both factors are constrained by CB rules.

A third factor means equally as much and suffers no restrictions. That's the *sensitivity* of the receiving portion of your transceiver.

Radio waves weaken as they travel. Signals picked up by any CB antenna from stations or mobile units not close by can barely be measured—just a few millionths of a volt (microvolts). These tiny signals must be greatly amplified by your receiver before the voice modulation they carry can be extracted. The better your receiver amplifies incoming CB signals, without adding tube or transistor noise, the more *sensitive* your receiver is said to be. You can even buy a preamplifier that builds up weak signals ahead of the receiver.

Receivers having a high sensitivity are expensive. They're rated by microvolts (μV) of signal required to operate the receiver at certain levels. A good receiver's sensitivity is better (lower) than 1 microvolt. Only a trained and equipped technician can accurately determine the sensitivity of your CB receiver.

Any discussion of distance among CBers inevitably turns to *skip*. This phenomenon got brief mention on pages 22 and 28. It's an oddity you should understand, even though you can't use it for CB (skip communications are illegal in the Citizens Band).

Radio waves at and near 27 MHz exhibit a tendency to travel straight. That means they don't follow the curvature of the earth, as some radio waves do. Instead, they head right out into space. EXCEPT, when the ionosphere has just the right mixture of electrical charges, 27-MHz signals bounce right back toward the ground. They hit the earth again some 600 to 1000 miles from the point of origin.

In rare ionospheric situations, they can bounce two, three, or even four times—ending up thousands of miles away. Weather causes some ionospheric shifts that create skip. The chief contributor is cosmic bombardment from outer space, generally heightened by sunspot activity (explosions on the sun).

Frequency has been mentioned. The word frequency has to do with the wave nature (form) of radio signals. You've seen the abbreviation *MHz*. These two terms relate to the channels your CB transceiver operates on.

Every radio station operates at some "carrier" frequency in the radio-wave spectrum. Radio waves have peaks and valleys just as water waves do, but they're electromagnetic fluctuations. Frequency indicates how many of these peaks and valleys a particular radio wave develops every second. One peak and one valley is one cycle. One cycle per second is called a *hertz*.

CB radio waves go through 27 million of these electromagnetic cycles each second. So, 27 million cycles every second equals 27 million hertz. The prefix *mega* stands for million; hence CB transmitters generate radio waves at around 27 megahertz, abbreviated 27 MHz.

Channel 1, for example, operates at a carrier frequency of 26.965 MHz. That's a wave with 26,965,000 peaks and valleys (cycles) each second. The transmitter sends out waves (signals) of that frequency; the receiver picks up waves coming from any other station on the same frequency.

CB Channel Frequencies

Channel	Frequency (MHz)	Channel	Frequency (MHz)
1	26.965	12	27.105
2	26.975	13	27.115
3	26.985	14	27.125
4	27.005	15	27.135
5	27.015	16	27.155
6	27.025	17	27.165
7	27.035	18	27.175
8	27.055	19	27.185
9	27.065	20	27.205
10	27.075	21	27.215
11	27.085	22	27.225
		23	27.255

FCC rules for Citizens Radio stipulate that the transmitter be *crystal*-controlled for stability. That means the radio waves are generated by an oscillator circuit whose frequency is determined by a quartz crystal. It would seem there should be one crystal for each transmitter frequency. (The quartz crystals fit inside tiny oblong metal cans.)

For accuracy and stability, the receiver should also be crystal-controlled. That suggests 23 more crystals, a total of 46. Very expensive. For that reason, early (and still some low-cost) transceivers utilized only four or six channels. You picked the few you wanted, and listened to and talked on no others.

Most modern CB transceivers incorporate an oscillator-circuit design called a *synthesizer*. This innovation allows 23 transmit and 23 receive channels to be synthesized from a comparatively few crystals with carefully worked out mathematical relationships. The advent of frequency synthesis explains why today you can buy a 23-channel transceiver at very reasonable cost.

Any high-frequency receiver (27 MHz is considered high frequency) takes careful designing to separate stations. That is, it must possess a special quality called *selectivity*. If its selectivity is poor, a CB receiver tuned to one channel might tend to pick up those on each side of it as well. That's because the channel frequencies are not very far apart (page 32).

Tuned circuits in one section give a receiver its selectivity. They're called i-f (intermediate frequency) circuits. If these circuits are designed well, the more the better—up to a point. But they're expensive, and not really efficient.

A better means is available, and some receivers use it. They incorporate a "mechanical" (actually piezoelectric) filter that effectively shuts out any channel or station above or below the one you select on the dial. You can't see the filter itself; it's inside a metal can along with a tuned circuit or two. But you can be assured of good selectivity if you know the receiver uses this kind of selectivity filter.

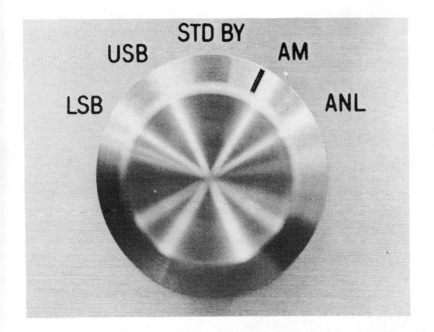

You'll encounter a knob on some CB transceivers with the letters AM and SSB—or perhaps AM, USB, and LSB. These letters refer to modes of operation.

Ordinary Citizens Band radios operate in a mode called *amplitude modulation.* That's AM (normally abbreviated as a-m in text). It refers to the way voice is added to the channel (carrier) frequency in the transmitter. The a-m process creates little "spillover" frequencies, called *sidebands,* right against the carrier frequency, one above and one below. They don't amount to much frequency-wise, but they are vitally important to the receiver. Sidebands contain the voice signal the receiver must recover.

SSB stands for *single-sideband.* This type of transmitter adds the voice in such a way that one sideband *and* the carrier are eliminated. Only the remaining sideband gets transmitted. The receiver is of special design too, and can recover voice from just one sideband.

The USB position of a transceiver mode switch sets the transceiver to send and receive the upper sideband, just above the carrier; LSB will send and receive the lower sideband. (More about ssb on page 41.)

An important part of your CB transceiver circuitry is the *modulation limiter*. You can't see it, nor do you need to. You should know what it does, or you may not understand some effects it introduces.

If you talk too close to the microphone, or too loudly, the nature of a-m and ssb is such that your voice could *overmodulate* the transmitter. Overmodulation creates multiple technical problems, and the FCC strictly forbids it. An automatic modulation-limiting circuit removes that worry from your mind. It clips the amount of voltage your voice can develop in the transmitter. No matter how loud you shout, the limiter keeps modulation from exceeding the legal maximum.

Your voice may become distorted if you speak too loudly into a microphone. A similar thing happens when you use some so-called "amplified" mikes. The transmitter can't accept a too-loud voice, so the limiter just squashes it down to what it should be. In the process, the excessive clipping can make your voice almost unintelligible to a listener.

Some CB units have a switch marked ANL. That stands for *automatic noise limiter.* Electrical radiation from the ignition system in some cars makes so much noise in the radio that communication with a weak or distant station becomes difficult. The ANL clips out noise of that type.

The noise could be coming from an automobile nearby, particularly at a base station with its antenna near the street. An electric motor, fluorescent light, or neon sign may generate noise impulses that the limiter can clip out or at least suppress.

Turn the noise limiter off when it's not needed. Communication with weak stations is clearer without it—provided there's no noise to overcome.

Want to keep the CB base station on, in case a unit calls in? Yet you're trying to read? What you need is something to cut out the noise of the receiver without having to turn down the volume control.

You have it—labeled *Squelch.* This knob shuts off the sound circuits of your receiver even with the volume control turned up. That is, it shuts them off when the receiver is not picking up a station. If a station suddenly comes on, the squelch circuits automatically open up and let the sound come through at whatever level you have the volume set.

Here's how to set it. Turn the Squelch all the way off, so you hear the "frying" noise of the receiver circuits with no station (turn to a channel no one is using). Then set the volume so you can hear the noise plainly but not loud enough to blast. Now turn up the Squelch knob until the sound suddenly stops. That's the squelch *threshold.*

Turn the Squelch knob just slightly further. That sets squelch a little deeper so random noise bursts don't open it falsely. If you want to hear only nearby transmitters, turn the knob further. The deeper you set it, the stronger the signal it takes to open up the sound circuits.

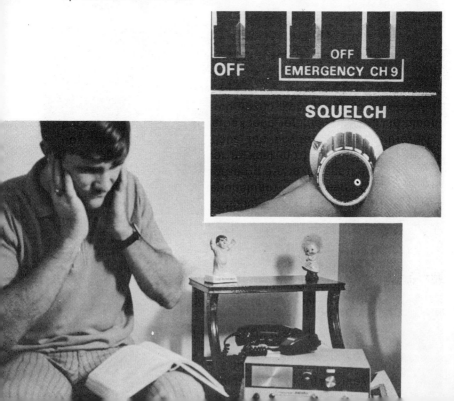

Chapter 3

Kinds of CB Equipment

A quick thumbing through any electronics catalog reveals how many different kinds of CB units you can pick from. And that's just from one supplier. Considering there are nearly two dozen major brands, you could have trouble making up your mind what to buy.

This chapter does not recommend specific brands. The illustrations exemplify various *types* of Citizens Band transceivers. You should be familiar with their differences and peculiarities, so your buying decision is at least informed. Then buy whatever best suits you and your purpose.

The ultimate in CB base stations sport modernistic console-design cabinets. A multitude of features make operation easy and accurate. The most expensive models offer ssb transmission/reception as well as a-m. Here is what some extraordinary features mean to you:

Compression-Type Modulation Limiter—full modulation at low voice levels and low-distortion limiting at high.
RF Gain Control—improves ssb reception over range of weak and strong signals. (A receiver can easily be overloaded by ssb.)
Five-Minute Automatic Timer—so you don't go past the time limit on station-to-station calls.
Built-In SWR Meter—to check antenna and cable; necessary only at time of installation.
Voice Lock—improves ssb reception by "holding" centered beside the incoming sideband.

Certain features, like the 24-hour digital clock, add value to a unit, yet they don't improve its performance. Others, such as a deep-acting noise blanker (limiter), are not even visible, yet they enhance operation. Study the specification lists of many units. Separate the frills from the nitty-gritty; then decide which model appeals most to you.

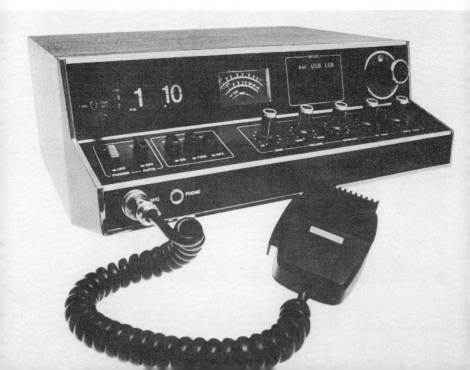

Single-sideband has distinct advantages for communications. Yet, be sure you understand this mode. Don't be misled that an am/ssb transceiver gives 69 channels. You talk and listen regularly in the a-m mode. Using the upper-sideband mode, you can talk and listen to other units operating in the same mode but only on the same 23 channels. The lower-sideband mode offers 23 more possibilities.

Just keep in mind that you still have only 23 channels to operate on. If you dial up channel 4, for example, your receiver picks up *all* signals on channel 4—whether a-m or ssb. The mode you've selected renders only its own signals clearly; a-m stations on the channel interfere the same as ever. And they hear you, though not intelligibly.

The advantage of ssb is its extra "sock." The transmitter can put more power *legally* into that one sideband. The result: your mobile units pick up a stronger signal, and distance range is increased. But the mobiles must be ssb units too.

While single-sideband transceivers hold the glamor of something different, plenty of top-grade transceivers remain strictly a-m. Fancy ones have 23-channel synthesizers, and some models cost little more than 6-channel versions. Remember, all your units must be ssb if you are to gain its advantages.

With input power limited to 5 watts, the only a-m transmitter specification that might impress you is output efficiency. If rf output is under 3 watts, the transmitter is inefficient. If more than 4 watts, it's illegal.

Concentrate on receiver sensitivity. Reread page 30 and grasp the full import. A receiver with sensitivity in the order of 0.5 microvolt (μV) pulls in weak signals from much farther than one with 1.0 μV sensitivity. The measurement standard should be 10 dB (S + N)/N for at least 2/3 of full audio output. Study those specifications. If you have the opportunity, let a technician verify receiver sensitivity in a unit you expect to buy.

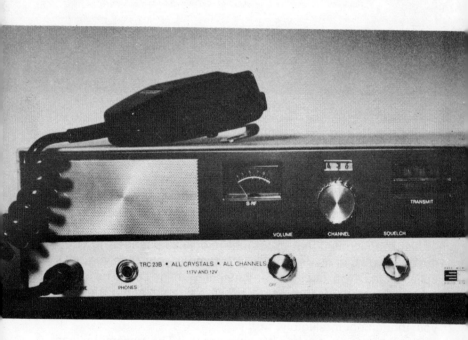

Your budget may be limited. You can acquire the fun and convenience of your own personal communications anyway. Just lower your sights a little. Settle for a simple six-channel transceiver.

Power output of its transmitter should be just as high and as far-reaching as from a 23-channel a-m unit. The receiver can be every bit as sensitive. But these are specifications you must always check before you buy low-cost equipment. You might be giving up some capability that is important to the purpose you have in mind. If all that's missing are unneeded frills, a low-cost model can be a bargain.

You can build your own 23-channel CB transceiver. Heath Company, Benton Harbor, MI 49022, sells a kit. The unit is all-transistor. To meet FCC regulations, frequency-determining portions of the transceiver are preassembled so you don't throw the unit off-frequency from lack of technical knowledge.

Nevertheless, before you put such a unit on the air, you should have the transmitter checked and aligned by a licensed transmitter technician. He'll check the frequency of each channel, transmitter modulation percentage, and the input and output power.

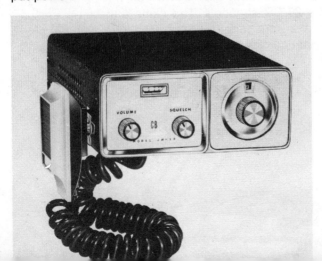

Some Citizens Band radios still use tubes, even though most manufacturers have changed to transistor designs. Tube models are fine for base stations because battery drain is no problem there.

You may find some excellent bargains if you go looking for a tube-type CB transceiver. Dealers know many buyers insist on solid-state equipment nowadays. A tube version he's carried on the shelf for a year or so might have a substantially marked-down price tag. Assuming the unit is new, it could give you years of perfectly acceptable service.

A few CB units have tunable receivers. For example, one six-channel model has an extra dial with which the operator can tune any of the 23 CB channels. The same unit has a crystal socket right on the front panel for the transmitter; you can plug in a proper crystal quickly for any additional channel. This isn't the handiest way to have 23 channels, but it's one method.

Recent innovations for the insides of a CB radio include *field-effect transistors* (FETs) and *integrated circuits* (ICs). They're added to solid-state models to give certain advantages.

A field-effect transistor offers special performance characteristics not available with conventional transistors. The FET, as it generally is used, improves receiver sensitivity without the transistor noise that ordinarily goes with high sensitivity. Some FETs are used for other purposes.

The integrated circuit is a modern miracle. One tiny chip, often no larger than a pinhead, contains the equivalent of several transistors and many associated parts. The drain on batteries becomes unbelievably small. The time may come when most of the circuitry in CB transceivers appears in the form of tiny ICs. But miniaturization is not a main purpose for ICs or FETs in CB radio. They appear in some large high-performance base stations.

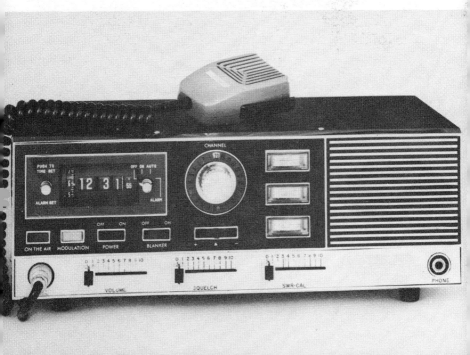

Having all 23 channels available may not be quite as important in your mobile units as in the base station. If you know which channels you'll use most for base-to-mobile talking, six could be plenty. That lets you find a channel clear enough for your communications—usually. (It's illegal as well as discourteous to butt in on someone else's conversation.)

Be sure your mobile has channel 9. That's the emergency channel for seeking help for yourself or another motorist. Pick four channels that are relatively unused in your area—if you can find that many. Then include one interstation channel (10 through 15) that is popular in your area, just in case channel 9 isn't commonly monitored. (See page 134 for teams that do monitor that channel.) In various parts of the country, channels 11 and 14 are popular for interstation calling.

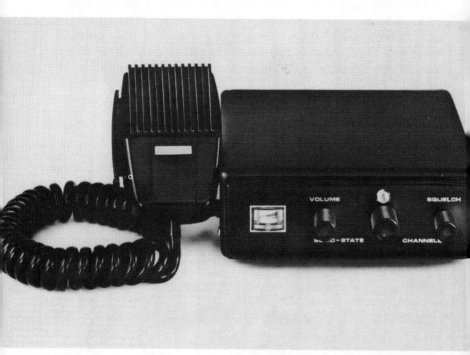

Mobile CB units operate on 12 volts dc. That's the nominal voltage of a car battery. Mobile units generally are very compact nowadays. They have no room for frills. With frequency synthesizers available, mobile transceivers with 23 channels have become common. They cost a bit more than those with 4 or 6 channels. You have to decide whether the added versatility is worth the extra cost to you.

A few mobile units include a pa (public address) feature. You mount a weatherproof speaker horn outside your car. Then, a switch on the CB unit routes what you say into the microphone, through the audio amplifiers of the radio, to that outside speaker.

Better check local ordinances before you hook this up. In some cities, outside loudspeakers on a car are legal only on emergency vehicles or by permit.

Base stations operate on 117 volts ac, the kind of voltage available at the wall outlets in your house. Mobile units use the 12-volt car battery since that's the only voltage ordinarily available there.

Certain transceivers can be operated either way. They have special power cords. Plug in one cord and the unit takes its power from 117 volts ac. The other cord connects to a 12-volt source, for mobile operation.

In the automobile, you generally use a handheld microphone. At home, the desktop mike seems more popular. The transmitter works fine with either kind of mike in either situation. Only the main voltage fed into the transceiver makes the difference between house and auto operation.

Even some fancy models you'd consider mainly for base-station use can be powered from the car battery. That opens additional possibilities. The unit can be used in the field for emergency communications—by rescue teams, for example. You have the operating advantages of a base station, yet independence from power-line voltage.

A 23-channel single-sideband base station, with all the extras you expect from a high-priced unit, can offer this multiple operating convenience. If you anticipate ever operating your CB system in a remote area, look over the specifications or examine the unit for dual-voltage operation. The cost shouldn't be much higher than for an ac-only model.

If you're an outdoorsman, you might want to add a couple of handheld CB transceivers to your equipment. They operate from self-contained batteries—sometimes dry cells, but often rechargeable nickel-cadmium batteries.

Some units operate only on two frequencies, but you can buy models for up to six channels. As with mobiles, be sure to install a crystal for channel 9, the emergency channel. That's the one most likely to be heard wherever you go. Some multi-channel portables come equipped with channel 9 crystals; you add whatever others you want.

Handheld CB receivers operate in the a-m mode. Be sure your am/ssb base or mobile unit (the one back at camp) stays set for a-m while you're away in the wilds.

Incidentally, a handheld CB unit counts as one of your quota of licensed units. Don't let those, plus your base, plus your mobiles, exceed the number of units stated on your license (see page 79).

Don't confuse Part 15 transceivers with Citizens Band handheld units. They are not the same, by any means. (And it's illegal to use Part 15 transceivers, which require no license, to communicate with licensed CB transceivers of any type.)

The chief difference in the two types of walkie-talkies lies in performance. With suitable terrain and a handheld CB unit that puts out 3 or 4 watts, you might talk 5 or 10 miles—maybe more. Part 15 transceivers rarely reach more than a mile or so. Their output is limited by law to 0.1 watt (100 milliwatts).

Another difference is frequency control. Part 15 transceivers may use just about any frequency in the vicinity of 27 MHz. CB transceivers, walkie-talkie or otherwise, must be crystal-controlled precisely on CB channels. You'll find this makes a considerable difference when you compare operation of the two types. Part 15 transceivers work best in pairs that have been tuned "to each other." A CB portable works with any other CB station that's on the channel(s).

The chief accessory for your Citizens Radio station, other than the antenna, is a microphone. Some kind of mike, usually handheld, comes with your CB transceiver. You might want one that sounds better, or is easier to use, or stands on the desktop.

Your CB dealer has display models you can get the feel of. But standing holding a tabletop microphone doesn't give the same perspective as having it in front of you at your base station. You might be fooled. When at all possible, don't buy a mike like this without an opportunity to sit down with it—more or less in an operating position.

Amplified microphones help if you want to sit at some distance from the mike. Or, if you want to lock the mike on and walk around the room as you talk, amplification helps. But the modulation limiter (page 36) prevents a microphone from adding anything to your modulation—*unless* the mike has a compression/expansion type of amplifier. That innovation keeps transmitter modulation close to maximum even when you lower your voice level occasionally.

For ordinary communications from an automobile, the usual handheld microphone serves well.

For noisy situations, such as in a truck or factory, a noise-canceling microphone works wonders. You hold it close to your mouth for talking, and the mike eliminates most of the noise that originates more than a few inches away.

Chapter 4

Antennas That "Reach Out"

You already know that distance with CB communications depends a lot on antenna design. As pages 19 and 20 explain, FCC rules limit both transmitter power and antenna height. You need a "gain" type of antenna to extend range farther than the height of your house allows (or a tree in your yard, if you put the antenna in the top of that).

CB's simplest base-station antenna is the ground plane. Its vertical "radiator" stands approximately 102 inches tall. That's just under one-quarter wavelength at CB frequencies of 27 MHz.

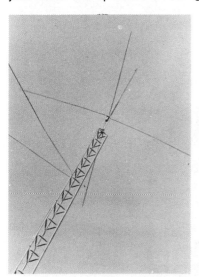

(Wavelength relates to frequency and to velocity of radio waves. Velocity is constant—300 million meters per second. The length of CB waves can be calculated by dividing 300 million by 27 million. The answer will be in meters.)

Dipole antennas (two separated rods) oriented vertically have characteristics similar in some ways to ground-plane types. The top rod forms the "radiator" for CB radio waves; the lower half is the "ground side" of the dipole. A dipole should be one-half wavelength at its operating frequency. For CB, that would be 17 feet tall—possible, but impractical.

To shorten the length, a device called the *gamma match* connects to a solid rod that looks like a dipole but is shorter. A gamma-matched CB radiator can be less than 11 feet long. Its radiation characteristics are similar to those of a ground-plane antenna. Waves radiate in all directions about equally, if the rod is vertical.

But designers add "gain" to the gamma-matched rod by altering its radiation pattern. Reflector rods behind the driven rod (radiator) reinforce radiation in the opposite direction. A slightly shorter rod in front aids this reinforcement. The array forms a *beam* antenna. It radiates strongly in one direction—toward the front (away from you, in the picture). The radiation, being thus concentrated, is much stronger than if it were omnidirectional. Your transmitter's CB signal reaches out much farther.

If one beam is good, two must be better. CB operators who need extra distance for their personal communications sometimes stack two beam antennas together on the same mounting. This requires a special "phasing" harness so the two antennas can be connected to the transceiver by one cable. Otherwise, two beams could be worse than one; they'd tend to cancel rather than reinforce each other. The resulting array is called a *stacked beam.*

A well-designed beam, adjusted properly, multiplies the effective radiated power (erp) of a CB transmitter. That's not to say the transmitter puts out any more signal. It means the antenna makes more effective use of the signal.

Any antenna reciprocates its transmission characteristic for reception. A beam is highly sensitive in its forward direction, but receives and transmits almost nothing at the sides or back. That means your mobile units must all be in one direction for efficient communications with the base station. Or, you can buy an electric rotator to swing the antenna beam to whatever direction you need at a particular moment.

Even the simple vertical ground-plane antenna can be improved upon. There is, for example, a five-eighths wavelength version. Its radiating element, the vertical rod, stands taller than the quarter-wave radiator. The rod must be sturdier.

The *radials,* as those bottom rods are called, have greater lengths too. You'll hear this one called the "five-eighths-waver" by CBers. Most have small *top-loading* circles or squares that shorten the physical dimension of the radiator slightly.

Another version of the quarter-wave CB antenna is shown at the left. It stands more than 18 feet tall. The design is termed *coaxial*. Instead of ground-plane radials below the radiator, a concentric *skirt* performs the same job. In fact, this coaxial construction makes the antenna considerably more efficient than a simple ground-plane type.

The lack of radials makes this omnidirectional antenna ideal for mounting where space is at a premium—for example, on a boat. The height demands a solid mount with a brace part way up the skirt portion of the mast. This antenna works well at home, too.

The antenna at the right is another one that needs very little surrounding space. Its bottom is *ring-fed,* a unique development. No ground-plane radials are necessary. This design more than doubles the effective radiated power of your base station in all directions, compared to a ground-plane antenna. This unusual design goes by the name of "Ringo."

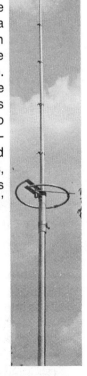

This antenna resembles no other you'll see. The vertical element is *top-loaded*. That term means that the shape and size of those rods at the top make possible a considerable shorter vertical rod without sacrificing efficiency. In fact, they improve the antenna's efficiency.

The peculiar-looking skirt contributes to radiation gain too. At the same time, it doesn't take up nearly the space ground-plane radials do. The manufacturer calls this omnidirectional model the Astro Plane.

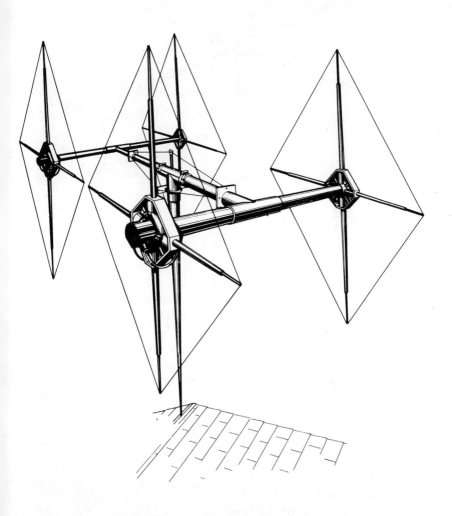

An antenna with rods crossed like this and metal wire connecting all the tips goes by the generic name of *quad* antenna. The one pictured here has special coupling devices that set it apart from other quad designs. The manufacturer calls this a PDL antenna.

What you see here is actually two antennas, connected together by a special phasing harness. This, you probably recall from page 55, is called *stacking*. Hence, the complete name for this high-gain (and directional) design is Stacked PDL.

The illustration gives you some notion of the relative size of a gamma-matched beam antenna. The black wire connected at the matching device (sometimes unimaginatively called the "matchbox") on the antenna is the *transmission line* or *lead-in*. It connects antenna and transceiver.

For most CB purposes, a quarter-inch line bearing the designation RG-58/U does the job. The cable construction is coaxial, giving rise to the nickname "coax" (pronounced COH-axe). An inner wire, of stiff copper and steel, has insulation surrounding it. Polyethylene foam makes the best insulation inside coax. On top of the foam coating comes a braided-wire metallic shield. The black outside covering is usually vinyl.

This coaxial cable has a special characteristic called its *impedance*. For CB, the characteristic impedance is 52 ohms. Unless transmitter, coaxial cable, and antenna all "match" this impedance, transmission and reception suffer. (Don't buy 72-ohm coaxial cable; it looks the same but is no good for CB.)

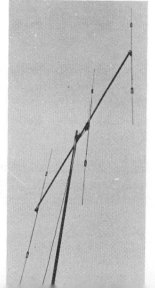

Antennas for mobile CB rigs have efficiency problems all their own. One of the best mobile antennas is a 102-inch whip antenna on a spring. It can be mounted in several ways, but a bumper clamp seems the most common.

At 27 MHz, the CB frequency range, the size of most cars doesn't offer a large enough "ground plane" for antenna effectiveness. That would be true even with the whip mounted in the center of the roof. The effect is even worse with a bumper mount. In fact, this kind of antenna exhibits directionality—across the corner of the car opposite the mount. Directionality is unhandy for a mobile. The bumper-mounted long whip has begun to fade from the CB scene.

The radiating element of any antenna can be shortened physically if it is "lengthened" again electronically. One *loading* method is with a coil.

This short whip resonates almost perfectly at CB frequencies. Yet, it and its small mounting spring are only about 3 feet long. A coil of wire inside the cylinder at the bottom adds the length needed for matching the CB transmitter electronically. The resulting antenna exhibits only slightly less efficiency than a long whip.

It offers one special advantage. The whole antenna is short enough to mount in a practical position on the car. The higher the better, and the roof would be best. But the trunk-lid center works fine. This mounting avoids some of the directionality associated with bumper-mounted whips.

The loading coil of a mobile antenna like that on the facing page causes a mild signal loss. Some CB antennas for mobile units reduce that effect by putting the loading coil up in the middle of the antenna rod or at the top. These are called *center-loaded* and *top-loaded* whips, respectively.

Some designers found that loading could be distributed along the length of the vertical rod. They made an antenna called *helical,* with loading wire wound around the entire length of the vertical element and then covered with a thin coating of plastic.

That idea evolved into the top-loaded fiberglass model you see in this photo. A *distributed-load* coil winds around the upper portion of the vertical rod. A fiberglass covering is flexible yet virtually impervious to damage. It keeps the loading coil fixed in place and protects the whole assembly from rain, snow, ice, and bugs.

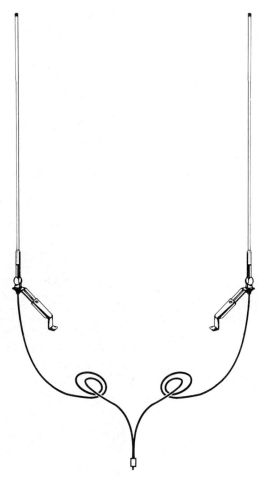

For distance, a mobile CB unit needs all the help it can get. Utilizing the principle of stacking, one antenna manufacturer came up with this co-phased dual array.

Identical whips, with distributed top loading, mount on each side of the car. The mountings fit into the trunk groove (next page) and leave no visible holes when removed.

The secret of using two antennas like this is a carefully designed *phasing* harness. Just as when beams are stacked, without proper phasing the antennas would "fight" each other rather than aid. Users realize a definite improvement in distance from this antenna array over an ordinary single-whip installation.

Cutting holes in a car body for antenna mounting can be costly. At least, filling them in when you get ready to trade the car gets expensive—for someone. The trunk-groove mount hides the mounting holes so they don't need to be filled in. That's why it has become rather popular. Pages 112 through 115 show how to install a mount like this.

The magnetic mount is another answer to the hole-cutting problem. This works fairly well for short CB whips, preferably those with loading coils at the bottom. A top-heavy whip, or one that catches too much wind, might pull the mount loose. You won't lose the antenna, because the coaxial lead-in keeps it with the car. But the antenna could scratch up the paint as it flops around.

If you buy a magnetic mount, make sure it's one with a powerful magnet. And don't let it get banged up; that weakens magnets.

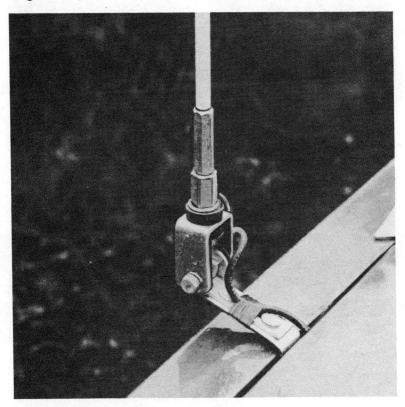

Switching between transmitter and receiver takes place inside the unit, so one antenna serves both. The coaxial lead-in cable from the antenna plugs into a connector jack on the rear of the transceiver. Three connector types have been used. Plugs for them, which fit on the transceiver end of the coax, are (clockwise from bottom):

Motorola pin-type—mostly on inexpensive units
RCA phono type—seldom used anymore
Amphenol type—by far the most popular

Each has its own special way of being installed. Pages 100 through 104 illustrate the steps. Your CB technician can install them if you'd rather not tackle it.

Chapter 5

Shopping for CB Gear

From what you've seen on earlier pages, you know by now what use you have for Citizens Band radio. Which leads you to the type of equipment you want and the antennas you'll need. The next step takes you shopping. Many companies sell by mail order. Look for their ads in electronics, CB, and do-it-yourself magazines. A postcard will usually bring their catalog. Many of these companies also have sales outlets in shopping centers, where you can see the merchandise.

Don't forget the Yellow Pages in your phone book. Look under "Radio Communication Equipment & Systems" first. Many companies listed there handle commercial two-way radio. Some sell or service CB radios too. Or, they can tell you who does. Check also under "Electronic Equipment & Supplies." Certain electronic distributors sell CB equipment to the public. If not, they know who the CB dealers are.

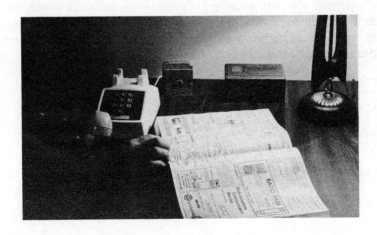

The thing to do, whatever kind of store you choose to shop in, is pick one that shows signs of specializing in CB or similar communications. An excellent example—a store in which an entire display area has been turned over to communications. You have an opportunity to examine just about every kind of radio equipment the company has to offer.

Turn on the transceiver you'd like to own. Fiddle with the knobs. A display like this one should be "live." That is, the equipment is hooked up and working. A demonstration tells you far more than a list of specifications. You want to know how the unit sounds. Compare units at various price levels. Do mobile models work as well as base units?

The salesmen in a store that has a special display usually have more-than-average knowledge of CB. They can advise you wisely and help you get the most for your communications dollar. When you buy equipment from someone who knows the field, you can just about be assured that you'll end up satisfied. He'll know the peculiarities of the equipment, and of your particular CB application.

Run through the controls of the transceiver—for both transmit and receive. How does the mike feel in your hand? If you want later to add a tabletop mike to the unit, can you? What kind of antenna connector (page 66) does the unit have? Does it fit the plug from your antenna? If you're a newcomer to Citizens Band radio, don't hesitate to ask the salesman for information that would help you operate his brand of equipment legitimately.

You gain a definite advantage when you can see and feel the operation of a CB unit before you spend your dollars. Even if you wind up ordering the unit from a catalog, looking and touching the real item beats viewing a picture.

Some stores specialize in nothing but communications. They are your safest bet when you know nothing at all about CB radio.

A company like this hires personnel experienced in more than one kind of personal communications. They generally have the technical expertise to help you solve your CB radio problems before they even arise. You can learn from them what to expect from whatever equipment you buy—transceiver, antenna, coaxial lead-in. They know of doo-dads that solve unusual problems, like an odd antenna mount or a beam with a special radiation pattern. More than likely, they keep a supply of license applications and can help you fill yours in correctly.

You'll find this type of firm listed in the Yellow Pages, usually under "Radio Communication Equipment & Systems."

What may someday be more important: a communications specialist knows where you can find someone to repair your CB equipment when something goes wrong. You should have your rig checked over at least once a year. Transmitter power, channel frequencies, modulation percentage, receiver sensitivity—all these need periodic attention, if only to keep your station legal.

Servicing facilities may be extremely modest. The best have expensive test instruments for measuring all those operating qualities accurately. For most transmitter work, a technician must be licensed by the FCC. Otherwise, he's not eligible to certify your transmitter as operating within legal limits. Go to the specialist if you find one; you'll save money in the long run.

While you're comparing prices and features of new or used CB equipment, give some thought to what you might do if something quit working. Ask that question of the salesman you are dealing with. He should be able to tell you specifically where you can get the unit repaired in case of trouble. If he can't, you really should shop some other stores.

. . . And don't fall for the old "they hardly ever go bad" routine. They do. When it happens to you, either you find a repair technician or you trade off the unit at a loss or you junk it. Unless the unit came to you at very low cost and you can get another the same way, the repairman is your better bargain.

So, however and wherever you buy CB equipment, do it with all these factors in mind. First, determine your particular need. Then compare features of all equipment that fits your need and budget. Finally, unless you already know where you can find servicing help, try to buy where the sales people know something about what they're selling . . . and that will include technical advice or the name of a place where you can find it.

You might want to know if a certain fancy and costly antenna would significantly aid your personal CB communications. Competent, impartial technical advice saves you dollars and aggravation. Find it before you plunge deeply into Citizens Radio.

Chapter 6

Getting Your CB License

Of all the radio communication services, Citizens Radio offers the simplest procedure for obtaining a license. The only prerequisites: that you are a United States citizen, 18 or older. With these qualifications, you have only to make application and pay the necessary fee.

Even the paper work has been simplified. A new application form makes that task easier. This chapter shows how to fill out the new form in such a way that you won't get it kicked back. All things going well, if you fill out like these pages indicate, you'll have your license in a month or less.

DO NOT operate a CB transmitter WITHOUT a license. Do not operate it "on someone else's license," unless you are employed by that someone or belong to same family *and* live in the same household as that someone. Otherwise, you operate illegally and could cause that person to lose his or her license also. Get your own license; it's so easy and inexpensive now that it's ridiculous not to.

Some manufacturers include a copy of the license application form with every new CB transceiver. If there's none with your unit, or you are buying a used one, drop a postcard to your nearest FCC field office. Ask for a copy of Bulletins 1001 and 1001h, and for FCC Form 505, *Application for Class C or D License in the Citizens Radio Service*. In most cases, you'll have everything you need within a week.

FCC FIELD OFFICES

439 U.S. Courthouse and Customhouse
113 St. Joseph Street
Mobile AL 36602
205-433-3581, Ext. 209

U.S. Post Office Building
Room G63
4th and G Street
P.O. Box 644
Anchorage AK 99510
907-272-1822

U.S. Courthouse
Room 1754
312 North Spring Street
Los Angeles CA 90012
213-688-3276

Fox Theatre Building
1245 Seventh Avenue
San Diego CA 92101

300 South Ferry Street
Terminal Island
San Pedro CA 90731
213-831-9281

323A Customhouse
555 Battery Street
San Francisco CA 94111
415-556-7700

504 New Customhouse
19th St. between California & Stout Sts.
Denver CO 80202
303-837-4054

Room 216
1919 M Street, N.W.
Washington DC 20554
202-632-7000

919 Federal Building
51 S.W. First Avenue
Miami FL 33130
305-350-5541

738 Federal Building
500 Zack Street
Tampa FL 33606
813-228-7711, Ext. 233

1602 Gas Light Tower
235 Peachtree Street, N.E.
Atlanta GA 30303
404-526-6381

238 Federal Office Bldg. & Courthouse
Bull and State Streets
P.O. Box 8004
Savannah GA 31402
912-232-4321, Ext. 320

502 Federal Building
P.O. Box 1021
Honolulu HI 96808
546-5640

37th Floor - Federal Bldg.
219 South Dearborn Street
Chicago IL 60604
312-353-5386

FCC FIELD OFFICES

829 Federal Building South
600 South Street
New Orleans LA 70130
504-527-2094

George M. Fallon Federal Bldg.
Room 819
31 Hopkins Plaza
Baltimore MD 21201
301-962-2727

1600 Customhouse
India & State Streets
Boston MA 02109
617-223-6608

1054 Federal Building
Washington Blvd. & LaFayette Street
Detroit MI 48226
313-226-6077

691 Federal Building
4th & Robert Streets
St. Paul MN 55101
612-725-7819

1703 Federal Building
601 East 12th Street
Kansas City MO 64106
816-374-5526

905 Federal Building
111 W. Huron St. at Delaware Ave.
Buffalo NY 14202
716-842-3216

748 Federal Building
641 Washington Street
New York NY 10014
212-620-5745

314 Multnomah Building
319 S.W. Pine Street
Portland OR 97204
503-221-3097

1005 U.S. Customhouse
2nd & Chestnut Streets
Philadelphia PA 19106
215-597-4410

U.S. Post Office & Courthouse
Room 322 - 323
P.O. Box 2987
San Juan PR 00903
809-722-4562

323 Federal Building
300 Willow Street
Beaumont TX 77701
713-838-0271, Ext. 317

Federal Building-U.S. Courthouse
Room 13E7
1100 Commerce Street
Dallas TX 75202
214-749-3243

5636 Federal Building
515 Rusk Avenue
Houston TX 77002
713-226-4306

Military Circle
870 North Military Highway
Norfolk VA 23502
703-420-5100

8012 Federal Office Building
909 First Avenue
Seattle WA 98104
206-442-7653

Your next step is to procure a copy of the FCC Rules and Regulations that apply to the Citizens Radio Service. You cannot get this from the FCC. You have to order it from the Government Printing Office. Fill in the coupon on the bottom of the new Form 505. Send that along with a check for $5.35 (not $3.50 as some older forms indicate) to the Superintendent of Documents, Government Printing Office, Washington DC 20402. What that brings you is Volume VI of the FCC Rules and Regulations. There's a lot of garbage in it you don't need, but also in it is Part 95, the laws that cover the Citizens Radio Service.

It is mandatory that you take this step. In signing your CB application, you stipulate and affirm that you either own or have ordered a copy of these Rules and Regulations. You are required to maintain a copy at the place your license hangs.

READ Part 95. It contains all the rules you have to be concerned with. It tells you, in legalistic jargon but nevertheless understandably, exactly how you can and cannot use your CB radio legally. You'll find the rules make pretty good sense. Nobody needs the yo-yos who clutter up CB channels with illegal inanities. CB radio comprises a valuable tool. Part 95 lays out operating procedures that assure everyone who needs it the opportunity to take advantage of CB radio.

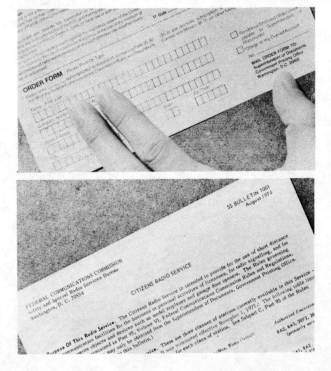

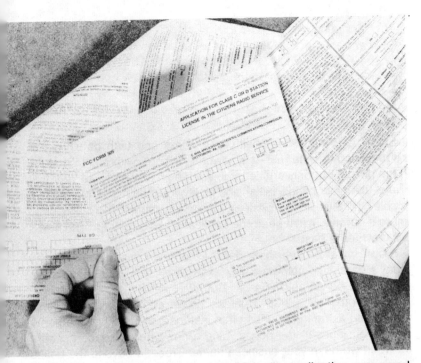

Here you see the new Form 505 CB radio application compared with the old one. Obviously, the new one-sheet form is far easier to fill out than the six-page (with instructions) old one. The next few pages reveal mistakes that are commonly made. Avoid those and you can rest easy that your license will be back from the FCC in record time.

The form has seventeen steps. Here's what to do with each of them.

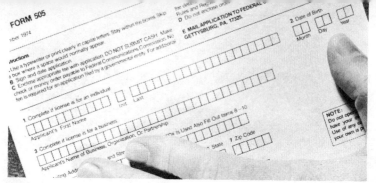

Steps 1, 2, and 3 deal with your name and birth date. As the instructions stipulate, use a typewriter or print, with capital letters, in the boxes. Put only one letter in each box. In step 1, put your first name, then your middle initial, and finally your last name. Make them legible, because if the clerks at the Gettysburg FCC office can't figure it out they'll just send the application back to you. That's why a typewriter is best.

Step 2 must be your birth date. A common mistake is to put in today's date. Use numerals. For example, July 4, 1932 should appear in the boxes as 07 04 32. Got it?

Don't bother with step 3 unless you're applying as a business rather than as an individual. In that instance, print or type the name of your business in the boxes. Skip a box when a space appears between words in your business name. Proceed beyond the boxes if necessary, but try to keep letter spacing the same.

Steps 4 through 7 call for a mailing address. If your application is in the name of your business, use your office mailing address. Be sure you include the zipcode. Your post office can tell you what it is if you don't remember.

Steps 8, 9, and 10 you fill in only if your mailing address is not a street address or not where your base will be. The FCC wants to know where your principal transmitter will be located. That's where your license should be posted. If you don't have a base station, give the physical location or address where you live. Or, if the application is for a business, list the main office. When you live in a rural location, give road coordinates or even latitude and longitude (ask the county surveyor's office to help you with these).

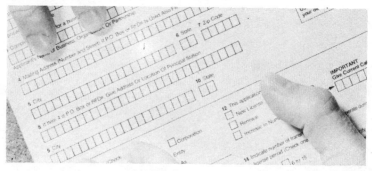

Step 11: Merely check whichever box describes your application. You can apply as an individual, even if you own and run a business. But if you apply as a business, be sure you have put the business name in the boxes of step 3.

Step 12: You use the same form for renewing a CB license to expire. In that instance, print your present call sign in the boxes provided at the right. If the old license has already expired, treat this as a New License application. You can use Form 505 to increase the number of transmitters for your present license. If you're increasing to more than 15 transmitters, you'll have to attach a statement saying why you need so many. Be sure to include your call sign for this too. If your application is for a new license, you need only put a check in the New License box.

Step 13: Check the Class D box. The other one is for a license to use radio-controlled model planes, boats, and cars. Voice communications requires a class-D license.

Step 14: Here's where you decide how many transmitters you plan to operate under your license. Include your family and any employees you will supply with CB radios. As in step 12, if you need more than 15 transmitters, tell the FCC why.

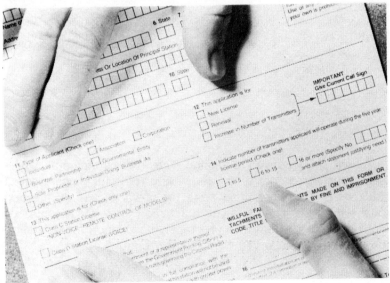

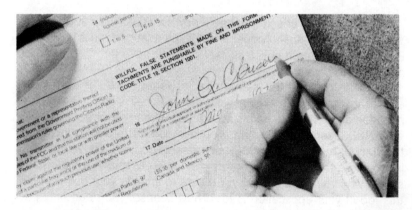

DON'T FORGET TO SIGN the application form (step 16). More applications are returned for this reason than any other.

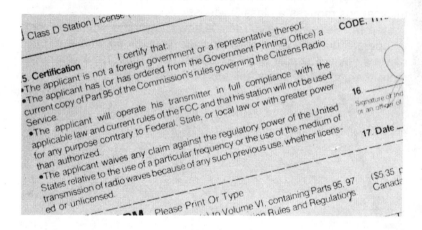

Also READ what you're signing. These four statements represent your agreement to the terms and rules for using class-D Citizens Radio transmitters. Note the power clause (at the end of the third statement). If you don't read these statements, it's your own fault when later you receive a citation for operating improperly because you "didn't know better."

Finally, don't forget to put the date beneath your signature.

There's nothing more to do except include the correct application fee. As this book goes to press, the fee is $4, having been reduced to that amount from $20. A postcard or phone call to any FCC Field Office (pages 76–77) will let you know in subsequent years whether the fee has changed.

Stuff the check and the completed, signed, and dated application form into an envelope, affix postage, and mail it to Federal Communications Commission, Gettysburg PA 17325. Do not send applications to field offices or to Washington. That merely slows down the processing by at least a week and sometimes longer.

You're not required to keep the license itself in your mobile. However, the transmitter in your vehicle must have some form of identification. Some operators carry a Xerox copy of their license. More convenient is a small tag called Form 452C, a design approved by the FCC.

Some manufacturers include a small ID sticker with each CB unit intended for mobile use. You can just as easily devise a tag of your own. It must contain the following information:

Name of licensee
Location of base (where license is)
Class of station (D)
Signature of licensee
License call sign

Stick this ID tag on your mobile unit, and that's proof enough that you have a license if any FCC inspector stops you to ask. However, don't fake it; he'll check you out and you might get a double fine (false information to a federal officer).

Your license is good for 5 years. The expiration date appears on the face of it. If for some reason you lose the license document itself, apply for a duplicate. Just drop a postcard to the nearest field office and they'll tell you the cost and procedure.

If you move, you're required to file a change-of-address with the FCC. On a postcard, give them your call sign, your old address, and the new address. Of course, don't forget your name. There is no charge for changing address.

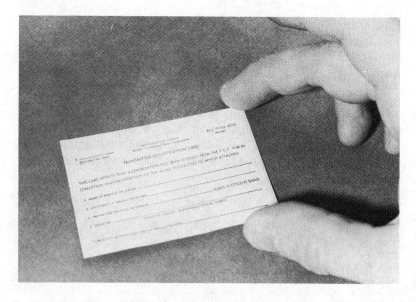

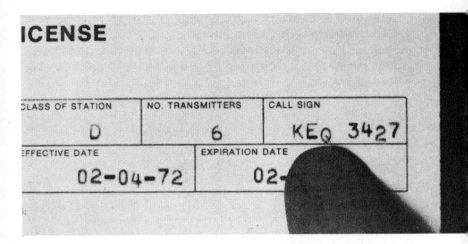

When your license arrives, usually two or three weeks after you apply, these are the most important bits of information on it. Your *call sign* (above) consists of three letters and four numerals. Memorize them. They're the only legal way to identify your station.

The expiration date appears just below your call sign. It's five years from the time the license was issued. Before your license expires, fill out another Form 505, mark it for renewal, make out another check (or money order) for whatever the license fee is then, and mail them. Keeping in mind that processing takes two or three weeks, apply for renewal about a month before expiration.

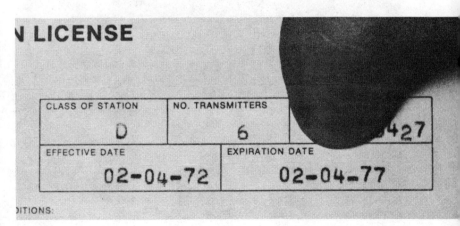

Post your license. Tape it or frame it, but put it up at your base transceiver. That's the law. If an FCC inspector takes a notion to look your station over by surprise some afternoon, your license looks nice hanging there where it belongs over your transceiver.

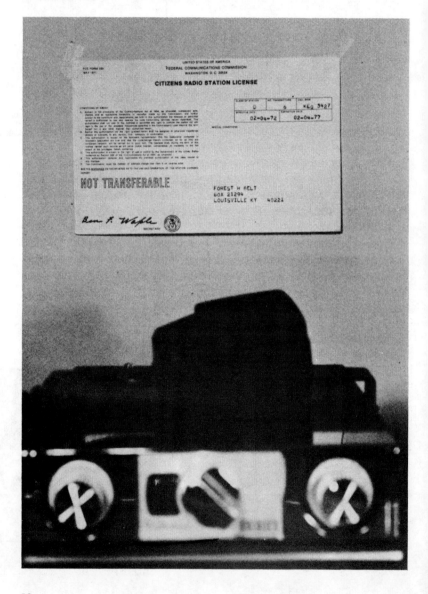

Chapter 7

Installing Your CB Base Station

The most ordinary Citizens Radio station can do a bangup operating job. Excellent performance does not depend entirely on the equipment. You've seen the reasons in earlier pages.

A kitchen table or cabinet holds a CB rig just as capably as a $250 desk. In an apartment or small house, a corner CB station may be the best you can manage. If it does what you want—bringing personal communications to your household—then who cares how simple the installation?

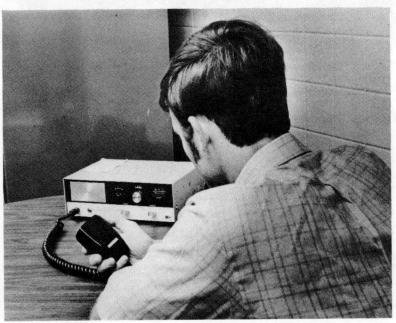

Of course, there's nothing wrong with a fancy station. But keep in mind that appearances come second at your CB base. Technically correct installation spells the chief difference between good performance or poor.

Appropriately, neatness usually accompanies technical competence. So if you pick the right expert to install your system, you'll probably have an installation that looks good *and* works well. If you're the "expert," you can accomplish acceptable results with a little foresight, knowledge, and caution.

Just don't sacrifice performance for appearance. With all its limitations, CB serves you well only when your base station functions at the peak of its potential.

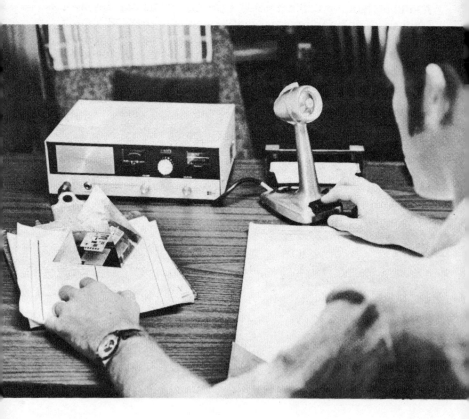

In your business, *utility* is the keyword. Success of day-to-day operations may well depend on how your CB base station works. It's well to reiterate: even a small CB transceiver can reach your mobile units as capably as a large expensive one. Any transceiver can legally deliver only 4 watts of rf power. Your antenna can extend only 20 feet above the highest point of your building.

Base-station antenna design could be limited by where your mobiles go. You can't use a directional beam very handily if your mobile vehicles head off in all directions. However, you might if your base is at one end of town.

An omnidirectional "gain" antenna (pages 56 and 57) might do. Or, if you send out only one mobile at a time, an electric rotator lets you swing a high-gain beam antenna to whatever direction you want. That keeps you turning the antenna for each trip the mobile makes; but the extra communications distance might be worth the trouble.

A Citizens Band base station takes the same basic hookup wherever you put the transceiver. Even out in the boondocks, you need these necessary components:

Transceiver
Microphone
Voltage supply for the transceiver
Transmission line (to antenna)
Antenna
Antenna support

Operating a CB station for several hours can be exhausting, so arrange for a solid table and a comfortable chair.

Interestingly, those are exactly the same requisites for your base station at home.

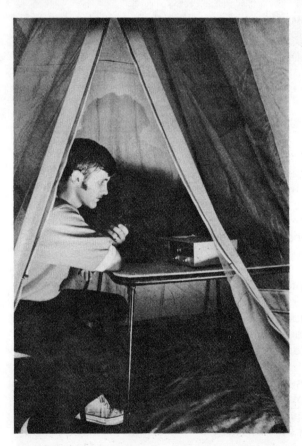

In the house, your fixed base station operates from 117 volts house power. You can run a 12-volt mobile transceiver at home from a car battery with a house-powered trickle charger connected to it. But that's messy and unhandy. If you have a 12-volt-only model, trade it in for a 117-volt ac-powered transceiver.

Really top-notch communications equipment nowadays uses three-wire plugs. The extra wire is a *ground wire* that connects to the round prong in the plug. The object is safety. The picture just above illustrates the kind of wall outlet this plug takes. Most CB units still use the cheaper two-wire plug, and you aren't forced to use this three-wire outlet. But if you're installing an outlet especially for your CB station, make it this kind. Sooner or later, all CB equipment will have three-wire plugs as it should.

Avoid this wiring "octopus." It can turn your cozy CB corner into a hothouse some night when it overheats.

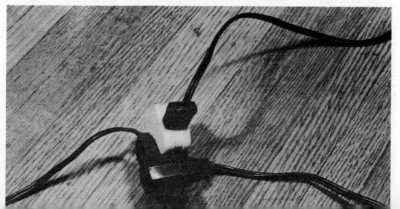

With a handheld microphone, you need a mike hanger. Pick a location that doesn't keep the mike cord stretched. Consider also whether you're righthanded or lefthanded. For hours of operating the station transceiver, you'll want to reduce every chance of fatigue. You shouldn't have to move from a comfortable sitting position to reach the mike, whether it's handheld or tabletop.

The transceiver and the antenna represent your two largest dollar investments. Unless you buy a kit, the transceiver comes ready to hook up and operate. The antenna, you have to assemble from a boxful of rods, screws, bolts, and clamps. You'll invest considerable time in the antenna installation if you do it yourself. Take care with it. Much of your station's performance depends on how accurately it's put together and adjusted.

READ THE INSTRUCTIONS completely before you even start slapping parts together. Identify any parts you don't recognize easily. Diagrams in the instruction sheet generally show each part by number. Don't be too embarrassed to write identification numbers on the metal rods and clamps; that could speed assembly once you start and assure getting the right parts in the right places.

Ordinarily, you assemble the elements first. Don't mix them up. Lay them out on the ground in their correct sequence as you put each one together. Label them any way you need to, but make sure you can mount them in their correct positions along the boom. Those positions are important. It's even important which end of an element points upward or downward (or to left or right, if you're assembling a horizontal antenna).

The antenna in these illustrations (a Mosley GD-3) is a three-element gamma-matched beam. The *gamma coupler,* a small black box, bolts right to the main boom. The connector for the coaxial transmission line points away from the driven element (the one that radiates the CB energy generated by your transmitter).

The *gamma bar* connects from the gamma coupler to one particular spot on the driven vertical element. The manufacturer has carefully calculated the spot and drilled a hole there for the gamma-bar mounting clamp. The gamma bar bends easily, so mount it and tighten its clamps with care. Don't let the clamps put a strain on it or the gamma bar will warp after you've raised the beam antenna into position.

You can buy coaxial transmission cable for CB that already has a connector plug at the antenna end. In fact, you can buy cable with plugs on both ends, but that complicates running the cable into the house neatly. The antenna end of the coax screws onto the gamma-box connector.

Mount the whole antenna on the pipe which will support it (the mast). Leaving some slack, and avoiding any pull on the coaxial connector, tape the coaxial cable to the mast. Wrap a few turns of tape around the mast first, then lay the cable over that and tape a 2- or 3-inch length. Don't be stingy with tape. Then, for extra strain relief, tape the cable tightly again an inch or two further down the pipe.

With the vertical beam clamped tightly to the supporting mast, check alignment of the elements. All should be in the same plane. Lying on the ground, the elements should be exactly parallel. If they're not, you can loosen certain clamps and adjust them. The instruction sheet tells how.

Finally, measure the length of each element. The instructions include a dimension diagram. Loosen and adjust any elements that are not exactly as prescribed. Then make sure all clamps and adjustments have been tightened and you're ready to raise the antenna and its support.

You can install lightning protection right in the coaxial lead-in. Two or three companies make this special coaxial lightning arrester especially for CB and communications use.

Install it right at the antenna connection on the transceiver. Then fasten the coaxial lead-in connector to the arrester. Run a wire, at least number 12 AWG, from the arrester screw to a good electrical ground rod—6 or 8 feet of copper rod driven into the ground. Put the ground rod as near your base transceiver as possible, to keep the ground wire short and direct.

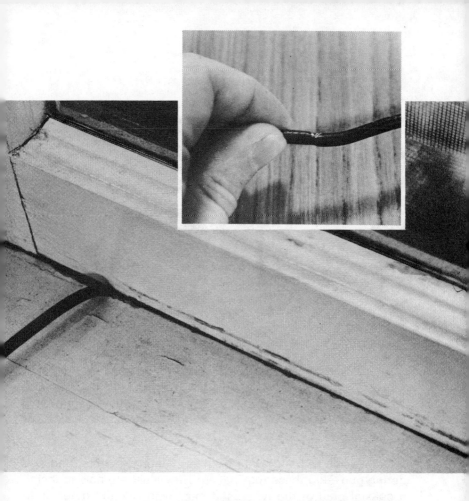

For any hole you drill in the framework of your house, be sure you aim the drill upward from the outside. That way, water can't run in. If you install a lightning arrester (opposite page), drill another hole about 6 inches from the lead-in entry. That's for the ground wire. Don't run it and the coax in the same hole.

Don't smash the coaxial cable any way, such as by clamping it under a window sash. Sharp bends in RG/U-type cables eventually foul the wire inside. That's a difficult fault to find, too; usually a technician can detect it with a reverse-reading wattmeter.

Unless you're handy with a knife, diagonal cutters, and soldering gun, skip these four pages. They illustrate how to install a coaxial plug of the type used most with CB and other communications transmission cable. It's not a job for the rank beginner, but it doesn't really take a technician. The prime criteria are carefulness and neatness of workmanship.

Cut the coaxial lead-in to leave some slack, but not a big coil of wire, behind the transceiver. Slip the screw-on ring and the size-adapter ferrule (for RG-58/U-size cable) onto the coax; they must be there before anything else is done.

With a knife that is not too sharp, saw through the outer vinyl sheath, the metallic braided shield, and the foam inner insulation of the cable. It pays to leave some extra cable if this is your first try. The trick is to slice *to* the inner wire without nicking it. Make this incision circular, all the way around the cable, about 2 inches from the end. Remove this 2-inch portion of sheath, shield, and insulation.

About one half-inch further back, circle the cable again with the knife. This time, cut through only the black vinyl outer sheath.

Peel back the braided shield. You can use the point of the knife to start it, but your fingers will do once you get the shield loosened. Unbraid the strands and fold them back toward the black vinyl sheath.

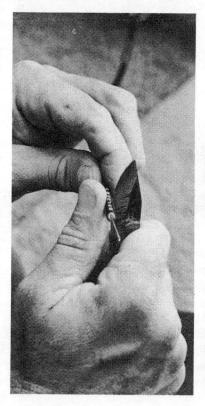 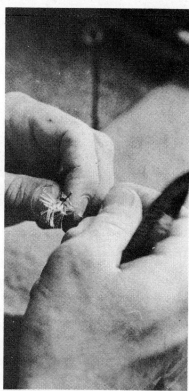

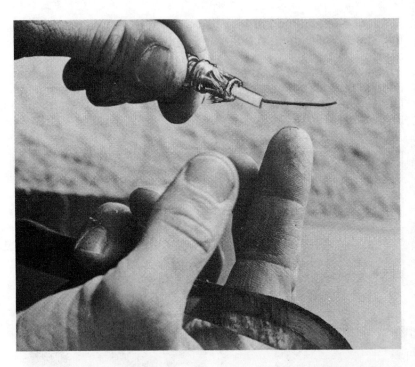

Slide the adapter ferrule down to the end of the black vinyl sheath. Work the shield strands back over the taper of the ferrule. Some of them will extend past the threads of the ferrule.

With a sharp-pointed pair of diagonal cutters, very carefully clip off whatever of these strands covers the threads of the ferrule. The shield-braid strands should cover only the smooth tapered end of the ferrule. "Comb" back and clip the strand ends until this part of the cable looks like the picture below.

This step requires meticulous care. No strand or "metal thread" must be left hanging loose. It can short out the plug later.

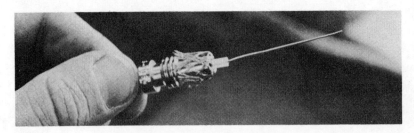

Now slip the main body of the connector onto the end of the cable. Run the bare center wire through the plug tip. Very slowly and carefully, so not to disturb the shield strands, screw the adapter ferrule into the main body of the connector. If it doesn't screw in all the way, some strands may still reach over the threads; back the ferrule out and trim the strands.

When you're sure the ferrule is smoothly seated and tight, heat the very tip of the plug and the center wire with a soldering gun. Melt some solder into the plug tip. Not too much. You don't want solder running down inside the plug.

You'll be able to see the shield strands through holes in the side of the plug body. Apply heat around *one* of the holes until you can run solder smoothly into the hole and onto the shield strands. Don't put solder into the other holes.

Now you can drop the outer screw-on shell down to the plug, and a few turns drops it into place over the plug tip. If you know how to use an ohmmeter, measure between the tip and the shell to be sure there's no short circuit.

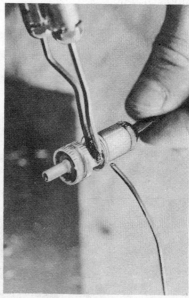

The photos show how Motorola pin-type (top) and RCA phono connectors fit onto coaxial cable. Only a few CB transceivers take these plugs in the antenna socket. But if yours is one of them, you should know how to prepare the cable end and solder on the plugs. Study the instructions on the preceding four pages, also; there are similarities in attaching all three kinds of plugs.

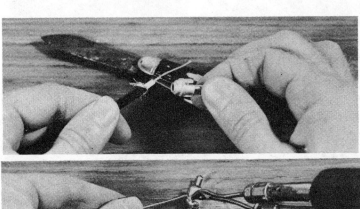

Chapter 8

Installing the Mobile Rig

The chief benefit of radio communications comes from mobility. You're not tied by telephone or telegraph wires to a fixed location. You can carry on radio conversations as you go about the day's errands in your car.

So . . . you'll undoubtedly install one or more mobile units. Or, you'll have it done. Generally speaking, a professional technician can do a better job than you can—at least in some respects. The money spent having the installation done by an expert might be saved in utility and convenience.

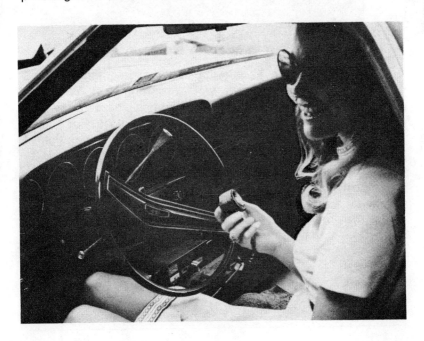

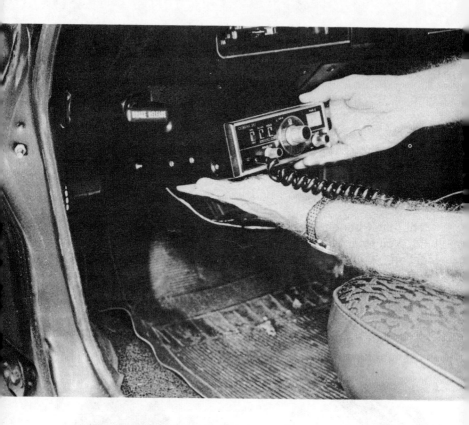

Figuring out where to put the unit in some modern cars proves as complicated as the work of mounting it. You can't put it where it gets bumped by—and bumps—passengers. The corners and the mike hangers snag women's hose. For safety, you have to be sure it doesn't interfere with your operation of brakes, clutch, parking brake, etc. And you shouldn't have to stretch to switch channels, set the squelch, adjust volume, or pick up the microphone. Not while you're driving; that's dangerous.

The pro has experience. He'll probably enlist your help in selecting a mounting position. But he's installed many units and knows the good and bad places. By trial and error, you can find them too.

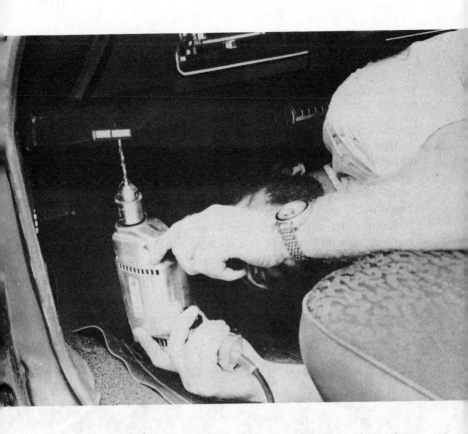

Be very sure before you start this step. Use the mounting bracket and a pencil to mark where the holes should be. Then, centerpunch the spots so the drill doesn't careen off and scar up the dash—or worse.

Feel above the lip of the dash before you start drilling. You can't afford to drill through something behind the dash. Even if you caught the drill before it hit something, the self-tapping mounting screws might extend far enough to rub some electrical wire bare.

A precaution: Use a drill with a three-wire cord and plug inserted into a grounding-type wall socket (the same kind as illustrated on page 91). No sense taking a chance of getting shocked.

Use self-tapping sheet-metal screws to attach the mounting bracket under the dash. Lockwashers with jagged edges (called grounding washers) under the heads of the screws make a solid electrical connection between bracket and dash. That helps avoid some sources of interference to the CB radio.

Tighten the screws, but don't overdo it. You can easily strip out the hole and have to use a larger screw. That might necessitate enlarging the hole in the mounting bracket—a lot of extra work.

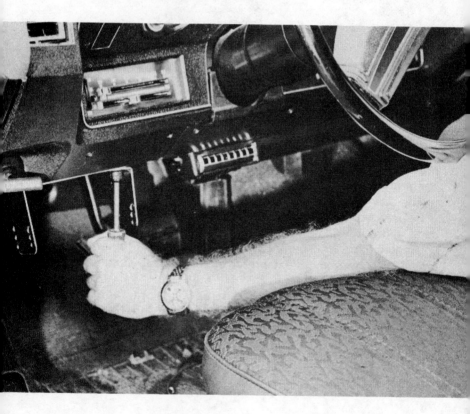

Grounding washers generally come with the screws or bolts that fasten the mobile transceiver into its mounting bracket. Use them. They continue the grounding you accomplish by installing grounding lockwashers on the bracket screws. Thorough grounding assures proper transmitter power as well as helping eliminate receiver interference.

Most mounting brackets have holes that permit varying the angle of the mobile unit. Try more than one position tentatively, with you sitting in your normal driving position. If the unit is near a passenger seat, have someone sit there too. There's always one most convenient angle. Try to find it.

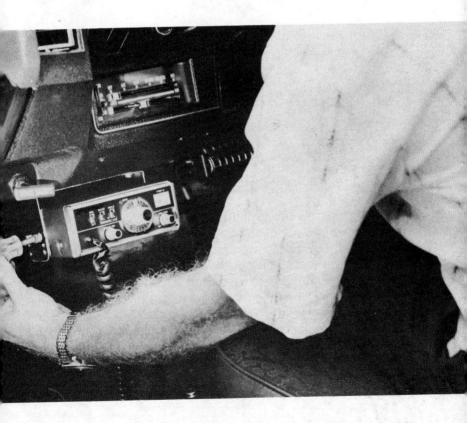

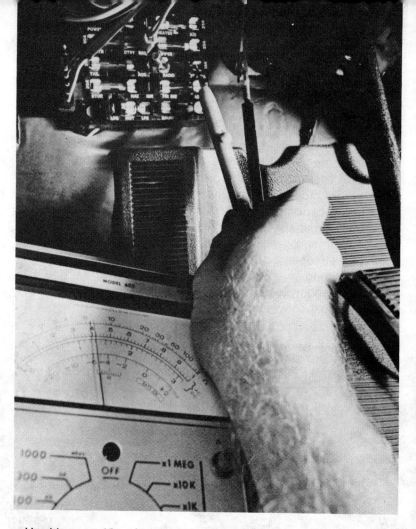

Hooking up 12 volts to the mobile transceiver can be fairly simple. These two pages illustrate an easy way.

The fuse block makes a good place to wire in. You want the unit to turn off when you cut the car's ignition. Go for a spot that's fed from the accessory portion of the ignition switch. It has voltage only when the switch is on.

Don't let the mobile transceiver put extra load on a fuse already in the block. Take a voltmeter and check which end of the fuse holder is "hot." You can do that by pulling the fuse out, turning the ignition switch on, and measuring at both ends of where the fuse clips. Only the "hot" end has voltage, as long as the fuse is out.

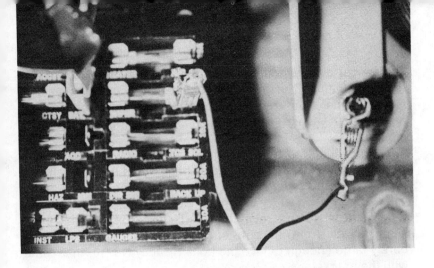

You need both a voltage and a ground connection. You get the voltage connection when you clip the transceiver's power lead (usually red) to the "hot" end of the fuse, as the photo illustrates. By clipping to that end, you don't pull current through the car's fuse. Yet, that's a handy place to put the clip you've attached to the transceiver wire.

The black transceiver wire goes to ground. You attach a clip to the end of that wire, too. Then hunt a nearby bolt or metal brace, wipe it shiny with sandpaper, and make the ground connection there. The photo shows the ground lead clipped to a protruding bolt.

The transceiver is protected from overload, even though you don't let its current come through the car's fuse. The transceiver has its own fuse. It's either in a holder on the back of the chassis or inside an "inline" fuseholder, such as the lower photo illustrates.

Most car owners don't want to drill a hole in the body for antenna mounting. The trunk-rim mount solves that dilemma. It leaves the car unmarred. The next few pages show the steps to a complete mobile antenna installation using this popular mounting.

Two small holes are all you have to drill. Hold the bracket in place, tight against the body (outer) edge of the trunk rim. If you don't mount the bracket close to the body there, the trunk lid will scrape when you try closing it. Mark the hole positions with a pencil. Centerpunch them to keep the drill from skidding even slightly away from the correct hole positions. Drill the holes barely large enough to take the bolts (usually No. 10) that come with the bracket.

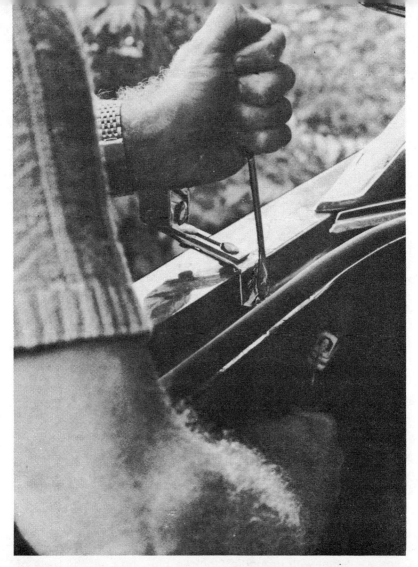

Hold the bracket in position, insert the two bolts, put nuts on them, and tighten with your fingers. Then check adjustment of the bracket, trying at the same time to close the trunk lid. Be sure the bracket allows free opening and closing of the lid, and that the lid doesn't get in the way of the bracket.

When you're sure the bracket is in the right position and adjustment, tighten the screws firmly. Use grounding-type lockwashers between the nuts and the underside of the trunk rim. Recheck the bracket and trunk lid for clearance once more before you consider this step complete.

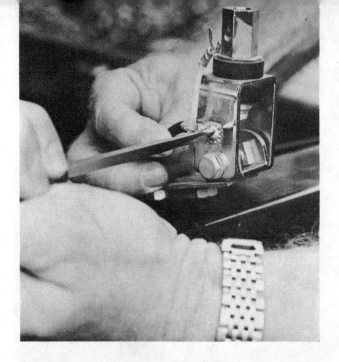

Prepare the end of the coaxial cable by stripping off about three inches of vinyl covering. Then pull the center wire and its insulation through the weave in the braided shield back near the start of the vinyl. That makes, in effect, a two-wire end for the coaxial cable.

To the braided shield, about one-half inch from where the inner wire comes through it, solder a terminal lug (see the photos). Cut off the leftover shield braid. That lug goes under a screw—with grounding lockwasher—down near the bottom of the bracket.

Strip off a quarter-inch of insulation from the tip of the center wire, and solder on another terminal lug. That lug goes under a screw—again with grounding lockwasher—up near the base of the antenna rod itself. (The rod hasn't been installed yet.)

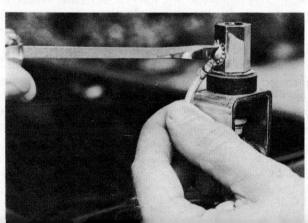

Orient the terminal lugs (opposite page) in such a way that the one on the shield supports the cable and aims the insulated center wire toward its attachment screw. Loosen the screws and retighten them, if need be, to get this orientation. You want the least strain. And you must avoid having the center wire bend sharply there at the braid. Strain or a sharp bend seriously shortens the life of this cable.

Then, tape the cable to the bracket body. Wrap a couple of turns of black plastic electrical tape around the bracket body just below the upper joint. Holding the cable in a way that leaves a tiny bit of slack between the taping point and the shield connection, wrap several turns of tape to hold the cable in that position. This relieves any pull the trunk lid might otherwise exert on the connections.

From there, route the cable under the trunk sealer lining and along the trunk rim for a short distance. Then let it drape over into the trunk. Unroll the length of cable. Twenty feet isn't too much for this purpose. Remember, it has to reach the transceiver under the dash, and not exactly by a direct route.

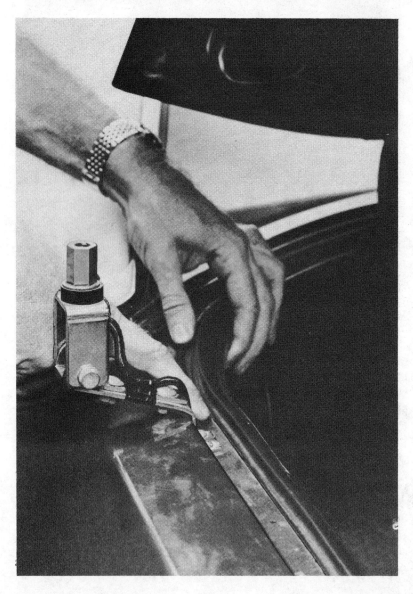

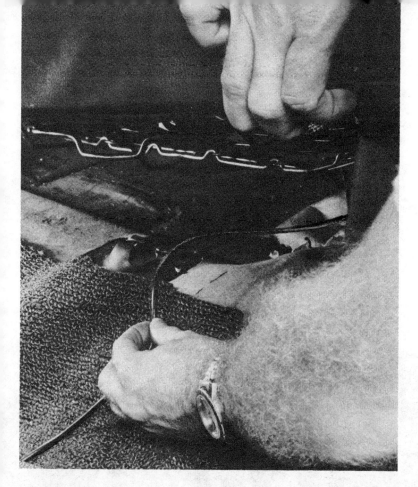

Leave reasonable slack between the trunk rim and the back of the rear seat. Dress (position and fasten) the cable close to the car body, so it doesn't flop around in the trunk. Then, from inside the trunk, poke the cable downward through one of the openings leading to the rear seat of the car.

If you aim the end downward and push it far enough, you can then raise up the rear seat cushion and grasp the end. Pull it gently. Recheck the cable in the trunk and don't let any kinks or sharp twists develop. Leave a little slack in the trunk but pull the rest through under the rear seat.

(If you don't know how to raise the rear seat cushion, feel along under its front edge. The front metal frame of the seat hooks under a small bracket. You can release the front edge by shoving the front of the seat downward, pushing toward the rear, then lifting upward.)

From the rear seat, the easy route for the antenna coaxial cable follows the floormats. Just lift up the rear one on the driver's side and slip the cable underneath. The cable then goes under the left side of the driver's seat, under the front mat, and up behind the dash near the accelerator pedal.

Tape the cable to a brace or to some of the wiring already under the dash, to support it. You don't want cable draping across from behind the mat to the transceiver. Your foot might get tangled in it. Besides possibly damaging the cable, careless dressing of the cable might put you into a dangerous driving situation. Sometimes you can tape the antenna cable and the battery and ground wires all together up under the dash. Find a support for them and fasten them there somehow.

If your car doesn't have floor mats, you can run the antenna cable under the carpeting. Just remove the four or five screws that hold the splash-guard trim plate. You can then lift up the carpet edge and slip the cable underneath. Screw the trim plate back in place.

The cable comes out about the same place as if you routed it under floor mats. Tie or tape it up under the dash as described on the page opposite.

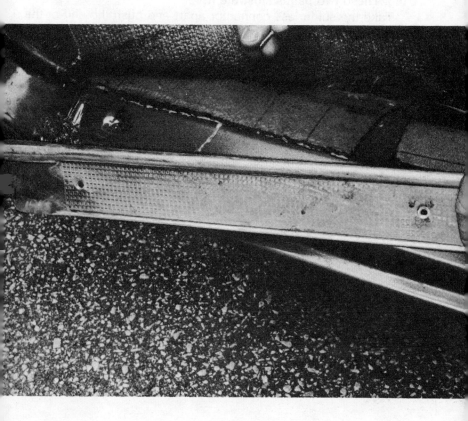

Buy your mobile antenna coax cable with a connector on one end already, if you can. That saves you the work of installing the connector. An installed connector will necessitate starting at the transceiver and running the cable backwards to the trunk and up to the antenna mount. Then when the length of cable is known, it can be cut and the connections to the antenna mount can be made, as previously described.

You can buy a connector that you can install without soldering. These two pages illustrate it.

Take the solderless connector apart and push the end of the coax through the body and coupling ring of the connector. Trim off 5/16 inch of the black vinyl sheath, being careful not to nick the strands of the braided metal shield. About 1/8 inch from the black vinyl covering, cut down through the braided shield *and* the inner insulation. But DO NOT nick the inner wire. Remove the 3/16 inch of braided shield and inner insulation you've cut loose.

If any braid strands extend more than 1/8 inch from the vinyl coating, trim them off with sharp diagonal cutters. Fan the strands backward over the vinyl. Every single strand should be back; any remaining forward could short out the connector.

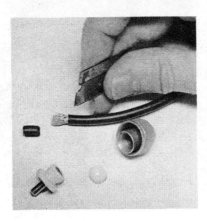
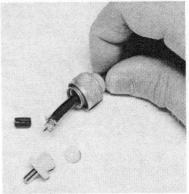

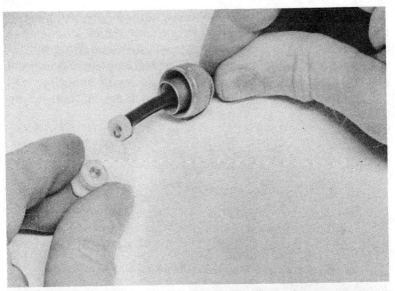

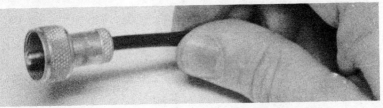

Slip the Teflon spacer down over the exposed insulation of the inner wire. The spacer should seat against the fanned-out braid strands and the outer sheath. Slip it outward just to make sure again that none of the strands get trapped underneath the spacer; then press it tightly back in place. With sharpnose pliers (or even tweezers), bend a curl in the center wire. It should be large enough to hold the spacer in place, yet too small to reach the outer edge of the spacer.

Slide the body and coupling ring down against the spacer. Screw the threaded Teflon pin-holder into the body of the solderless connector. The pin extends through the Teflon holder and contacts the spiral loop you made in the center wire. Screw the holder tightly into the connector body. That also holds the braided shield tightly in place behind the spacer. (When disconnecting this or any connector from the transceiver, never pull the cable; unscrew and handle only the connector.)

Finish off the mobile antenna job by screwing the antenna rod into its mount. You can leave it finger-tight and experience fine results with the transceiver. That facilitates removing it for going through an automatic car wash or entering a low garage. If neither of those factors applies, use a pair of wrenches and tighten the antenna securely. (You can see the finished job on page 63.)

The true finale comes when a licensed technician connects his inline wattmeter between the transceiver and the antenna. He can, first of all, touch up the output tuning to make sure your transmitter matches the cable and antenna. He'll also see that the transmitter puts out all it's capable of—up to the 4-watt legal maximum.

Then, with the reverse-reading function of his meter, he'll check the voltage standing-wave ratio (VSWR) of your antenna. If the antenna is adjustable, he can "tune" it for minimum VSWR and thus assure maximum radiation efficiency. That'll make sure you reach as far with that 4 watts as you possibly can.

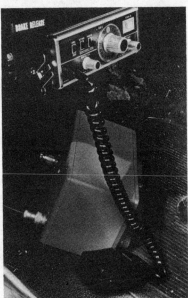

Here's something just about every do-it-yourself installer forgets. This is a transmitter identification sticker. FCC regulations require such an identification on every transceiver that isn't located where your license is posted. You should have an ident on each mobile unit. The same applies to handheld transceivers. (The transceiver at your office should have a legible photo copy of the license taped to it.)

You can get Form 452-C ident tags free from the nearest Federal Communications Commission field office. Addresses are on page 84.

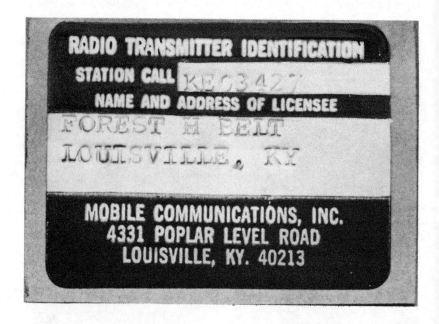

Chapter 9

Operating on the Citizens Band

Some of the earlier pages of this book have a sober tone. That's only because some thought, consideration, and work must precede your enjoyment of CB radio. But once you've got the license, bought equipment, and installed it well, you have almost clear sailing. You'll surprise yourself at the ways you'll find to use your personal communications system.

You're free to use CB in conducting your business. You have no obligation to operate your CB equipment for any purpose outside your own benefit and pleasure. You've already seen the minor limitations the FCC places on CB operation, and none of them need interfere with your making radio communications a consistent part of your household.

The next few pages deal with the actual fact of operating your radio on the Citizens Radio channels. Not everyone with a CB license knows or follows the most effective communications practices. Some operate thoughtlessly, some discourteously, and a few downright maliciously.

Newcomers to CB often are unaware of good operating practice. After you finish this chapter, you can run your CB rig as a competent communications operator.

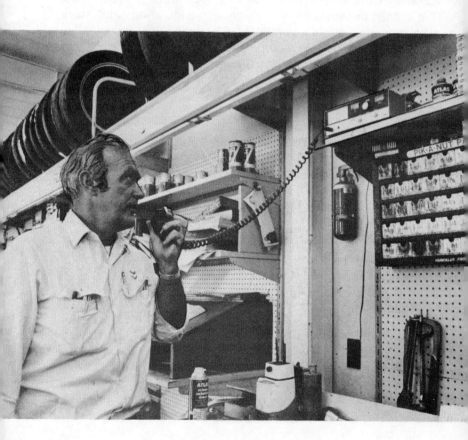

Considering FCC rules for Citizens Radio, you will be surprised at some things you hear on the air. Unfortunately, a few groups of smart-alec CB owners in every town hog certain channels. They chatter, giggle, insult each other, and make fools of themselves showing off. A few press their mike buttons down when others are talking, just to annoy. Such childish thoughtlessness spoils some of the usefulness CB offers sensible operators.

You can hope these inconsiderate gangs of CB delinquents don't rove the channels in your locality. Where they do, it's generally evenings. FCC inspectors periodically drive them into hiding, but they pop up again after the heat's off.

Your best recourse is to ignore their silly antics. Consider them ignorant, and refuse to engage in exchanges with them. If all the channels you could use are swamped, just turn off your rig and hope for the time when such operators grow up. (For the record, the operators in these two photos do not participate in the inanities they are depicting. But it does look juvenile, doesn't it?)

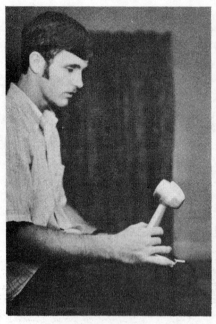 

Here are some operating techniques to improve your communications. The two different kinds of CB microphones require separate handling.

Handheld mikes are for close talking. Your voice sounds best to whoever's receiving when you hold the mike about two inches from your mouth. Don't talk directly into the face of the mike; your breath puffs raise distortion. Talk across the front, holding the mike at a slight angle.

A tabletop mike overloads easily if you talk close. Eighteen inches to two feet is a fair distance. Aim your talk directly at the mike, not across it. If you get reports of distortion, talk from a little farther away or a little softer. If you're in a room without carpeting or drapes and you get reports that your voice sounds like it's in the shower or in a barrel, talk a little softer and a little closer.

The worst operating habit of beginners is shouting. They unwittingly raise their voice to "project" it to the listener. If your transceiver and mike are any good, that's not necessary. Talk as if you were carrying on a conversation across the desk or table.

This combination S-meter and power meter can help you spot troubles in your rig before they become complicated. An incoming signal pushes the meter needle upward to a number that indicates comparative signal strength of the station. A signal that moves the needle to about S9 should sound clear—without receiver noise. Of course, interference from skip or other stations could make any signal noisy.

Watch the meter when you're transmitting. Once your rig has been installed and adjusted, the power meter reading should continue about the same. A dropoff in power suggests trouble in the transmitter, antenna, or lead-in.

Modulation (talking into the mike *modulates* the transmitter) wiggles the power needle slightly. Once you're familiar with how much movement your voice causes, you'll notice if that part of your transmitter starts growing weak. Some transceivers use a modulation light. This indicates just as well, once you're accustomed to it.

You should follow certain procedures to initiate communications with another station. The first step: select a channel that's legal—10 through 15, or channel 23. Be sure no one else is using it. Then lead off with your call sign. "KEQ-3427 calling KXX-9999."

Or, you can give the other station's call first. Repeat it a time or two. Then follow with your own. "KXX-9999. KXX-9999. This is KEQ-3427 calling KXX-9999."

Don't use a nickname or "handle" when initiating a call. You may be "Supercat" to the local CBers but the FCC knows you by call sign. Use it first and you won't get a citation for failing to identify your station. Then be Supercat all you want to. (Don't forget the rules about 5-minute interstation calls and about necessary communications.)

Call the other guy by whatever name he wants to be called. It's up to him to properly identify his station by call sign. The FCC has ways to track down CB users who fail to identify.

You can call other stations at random only if you're a mobile and need directions. "This is KEQ-3427 calling any station near downtown Hooplesburg. Come in please."

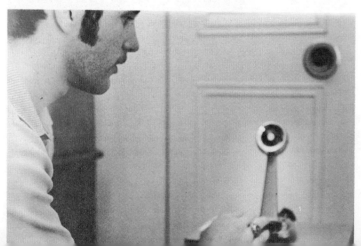

There's a correct way to call other units operating under your license. Give your call sign, identify which unit you are, and then stipulate which unit you're calling. "This is KEQ-3427 base calling unit 3." The FCC requires that you identify your station by call sign. You can use a name if you wish, just as long as you also identify by call sign.

The mobile unit operator answers, "This is unit 3; go ahead, base." No need for the call sign here. It was given at the start of communications. If the interchange continues more than 15 minutes, the base operator must repeat the call sign every 15 minutes or oftener.

A mobile unit operator initiates a call: "KEQ-3427 unit 3 to base." The base operator answers, "This is base, unit 3; go ahead."

Professional radio operators close every communication by repeating the call sign. "KEQ-3427 clear and standing by on 4." The mobile may also repeat, "Unit 3 clear and by on 4." If the conversation was obviously an ending one, the operator may say only "KEQ-3427."

Police and fire department radio operators years ago devised a code of numbers to speed "standard" communications and make them more accurate and understandable. It's called the *10-code*. Naturally, CBers have a similar code to simplify certain of their communications.

An example of its brevity: "KEQ-3427 base to unit 2. 10-20." "Unit 2. Fifth and Oak." "Unit 2, pickup at 1721 Whitmer. 10-33 please." "Unit 2 to base, 10-9." "Pickup at 1721 Whitmer, unit 2. 10-33." "10-4, base. 10-76." "KEQ-3427 clear." A lot of information was passed there, yet the whole exchange took less than a minute.

There's one trouble with 10-codes. Every CB book or magazine offers a different meaning for some numbers. The 10-code reproduced here comes from the Associated Police Communications Officers, Inc. and is used in the Citizens Radio Service by REACT monitors (page 134).

Code	Meaning	Code	Meaning	Code	Meaning
10-0	Caution	10-33	Emergency	10-64	Message for local delivery
10-1	Unable to copy — change location	10-34	Riot	10-65	New message assignment
10-2	Signals good	10-35	Major crime alert	10-66	Message cancellation
10-3	Stop transmitting	10-36	Correct time	10-67	Clear to read net message
10-4	Acknowledgement	10-37	Investigate suspicious vehicle	10-68	Dispatch information
10-5	Relay	10-41	Beginning tour of duty	10-69	Message received
10-6	Busy — Stand by unless urgent	10-42	Ending tour of duty	10-70	Fire alarm
10-7	Out of service (Give location and/or telephone number)	10-43	Information	10-71	Advise nature of fire (size, type & contents of building)
		10-44	Request permission to leave patrol for		
10-8	In service	10-45	Animal carcass in lane at	10-72	Report progress on fire
10-9	Repeat				
10-10	Fight in progress	10-46	Assist motorist	10-73	Smoke report
10-11	Dog case	10-47	Emergency road repairs needed	10-74	Negative
10-12	Stand by (stop)			10-75	In contact with
10-13	Weather and road report	10-48	Traffic standard needs repairs	10-76	En route
10-14	Report of prowler			10-77	ETA (Estimated time of arrival)
10-15	Civil disturbance	10-49	Traffic light out		
10-16	Domestic trouble	10-50	Accident — F, PI, PD	10-78	Need assistance
10-17	Meet complainant	10-51	Wrecker needed	10-79	Notify coroner
10-18	Complete assignment quickly	10-52	Ambulance needed	10-82	Reserve lodging
10-19	Return to	10-53	Road blocked	10-84	Are you going to meet
10-20	Location	10-54	Livestock on highway		
10-21	Call by telephone	10-55	Intoxicated driver	10-85	Delayed, due to
10-22	Disregard	10-56	Intoxicated pedestrian	10-87	Pick up checks for distribution
10-23	Arrived at scene	10-57	Hit & Run — F, PI, PD		
10-24	Assignment completed	10-58	Direct traffic	10-88	Advise phone No. to contact
10-25	Report in person to	10-59	Convoy or escort		
10-27	Drivers license information	10-60	Squad in vicinity	10-90	Bank alarm
10-28	Vehicle registration information			10-91	Unnecessary use of radio
10-29	Check records for wanted	10-61	Personnel in area		
10-30	Illegal use of radio	10-62	Reply to message	10-94	Drag racing
10-31	Crime in progress	10-63	Prepare to make written copy	10-96	Mental subject
10-32	Man with gun			10-98	Prison or jail break
				10-99	Records indicate wanted or stolen

You should be aware of an FCC dictum called *priority of communications*. Paragraph 95.85 of the FCC Rules and Regulations says:

"(a) All Citizens radio stations shall give priority to the emergency communications of other stations which involve the immediate safety of life of individuals and the immediate protection of property." This applies to all channels.

A note in paragraph 95.41 also lists this order of priorities in the use of channel 9, the emergency channel:

1. Communications relating to an existing situation dangerous to life or property, i.e., fire, automobile accident.
2. Communications relating to a potentially hazardous situation, i.e., car stalled in a dangerous place, lost child, boat out of gas.
3. Road assistance to a disabled vehicle on the highway or street.
4. Road and street directions.

If you're on any channel and someone has a priority message, you should clear and stay off until the priority communication has ended. Unless, of course, you're the handiest station to relay the message or render assistance.

The Citizens Band was created for *business* and *personal* use. CB and other radio communicating facilities can also be put to work during a genuine *emergency,* even ignoring many ordinary limitations. A fourth category of communications, *public service,* has developed into a major use of CB radio equipment.

Numerous small CB clubs operate locally to offer various public services. The largest national association of this nature is REACT. The name is an acronym for *Radio Emergency Associated Citizens Teams.* REACT grew from a small group of public-spirited CBers. Today, with national headquarters in Chicago, REACT boasts more than 1300 teams throughout the United States. General Motors Research Laboratories sponsors the organization, which is dedicated to emergency assistance.

REACT serves motorists, mainly. One activity includes monitoring channel 9, the emergency and assistance channel. In some localities, REACT teams listen to channel 9 at all times of the day and night.

Other REACT teams add other emergency and assistance activities. Several groups spend their weekends patrolling sections of interstate highways watching for motorists with car trouble. They stop and lend aid or perhaps radio for gas, a wrecker, tire service, or whatever.

If you have an urge to put your CB radio to work helping improve your community, consider joining REACT. Members help out with traffic to and from major community events. Some teams are trained in first aid and render valuable help anywhere a disaster, large or small, has overtaken their neighbors and fellow citizens.

The address for more information:

>REACT National Headquarters
>111 East Wacker Drive
>Chicago, IL 60601

Service-minded citizens can form their own associations for public service and use Citizens Radio as a communications medium. An example is the Kentucky Rescue Association (KRA). Like REACT, this group began small, mostly in Jefferson County (surrounding Louisville). Eventually, they united with similar groups in other counties. The larger group incorporated as the Kentucky Rescue Association and now has chapters statewide.

Emergency rescues are the KRA's stock-in-trade. Members stand ready to roll on missions at a moment's notice. Specialty squads are trained in underwater search and rescue. Extraction experts can take a car or truck apart in minutes to free a trapped victim. Some members are trained to excavate collapsed buildings. Cave specialists go in after lost spelunkers. Hundreds of members and friends can be mobilized for lost-child or lost-hiker searches. Members are trained in first aid.

Citizens Band radio takes care of most communication needs for Kentucky Rescue Association members. The automobiles of most members have CB. Likewise their homes. For field work, some teams carry walkie-talkies—the licensed kind.

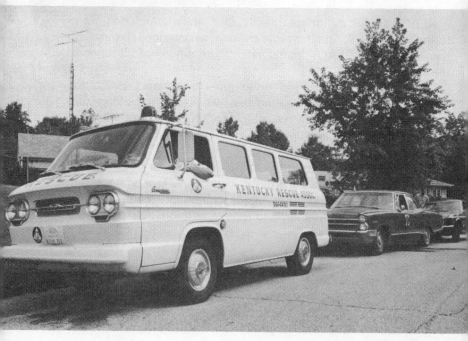

Being headquartered along the Ohio River, the KRA spends considerable time and effort in water rescues and recoveries. CB-equipped boats, all owned by members, carry underwater and sweep crews.

CB radio serves a unique function in circumstances like this. It's the only radio communications service that can be operated in boats, in cars, in planes, and at home—all under one set of FCC rules. Coordinating an areawide, multifacility search would present a tough communications problem without the Citizens Radio Service.

You can't credit CB with formation of a professional outfit like the Kentucky Rescue Association. But CB does make its operations handier and more efficient. You can put CB to work helping your club group do its share to make the community a safer place to live.

A closing word on your responsibilities as a CB licensee. You are held accountable for any communications your licensed units produce. Any interference by one of your transmitters is your worry. But only up to a point.

Take television interference, for example (TVI, it's called). Suppose someone complains that your CB rig causes lines (maybe not as severe as this photo) in their TV set. It's possible. But if the transmitter doesn't interfere with your own TV picture, the fault might not be yours.

However, make sure. Have a qualified, licensed technician check your transmitter. If he certifies that its harmonic radiation and power output do not exceed legal limits and that all your transmitter channels are on-frequency, the fault lies with the TV receiver.

A device called a high-pass filter—or a 27-MHz trap—can be installed at the TV set's antenna terminals. That'll knock out most CB interference. But it's a job for the TV owner's repair technician, not you.

Chapter 10

Adjustments and Maintenance

If that hand is yours, you're taking chances—unless you're a qualified technician. Certain voltage points inside this and any ac-powered CB transceiver are killers.

If you don't care about yourself, you're still taking a chance of damaging the rig. Doodling around inside your CB transceiver gives very little possibility of curing any ills it might have. The likelihood is that you'll only add to the repairs that will have to be made by some technician. Don't complain when the bill is higher than it would have been if you'd left the lid on.

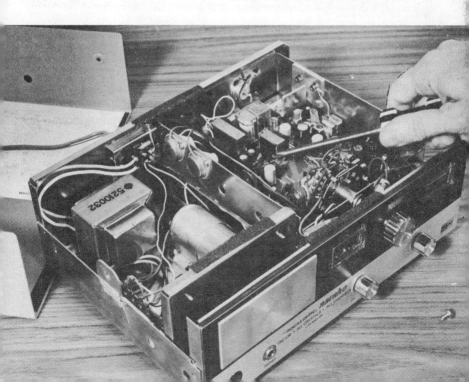

You know your transceiver comprises two sections: transmitter and receiver. If the unit operates from 117 volts ac (a wall outlet), there's a third section—the power supply.

If you own an older CB transceiver, one that uses tubes instead of transistors, you can resort to this repair method. There are a couple of things to remember. Do-it-yourself testers don't reveal some defects that foul up a tube for communications. Also, after you've replaced a tube, circuits related to it may need realignment.

Yet, this small chance to save a dollar or two may be worth considering. Be sure you know exactly which socket each tube comes out of. Draw a diagram. Put the tubes back in the same place. Unless you have the instruments and knowledge for it, DON'T try adjusting any coils or trimmer screws in the transceiver. You could ruin performance, and even make the transmitter operate illegally.

If you own a good rf signal generator *and* know how to use it, you can repeak alignment of the receiver section if performance seems to have fallen off. A nonmetal alignment tool is preferable to the metallic screwdriver pictured below. You can approximate a professional alignment job *if* you know how.

If you don't, then stay out of these coils. They have a deep effect on how well your receiver picks up incoming signals. Mess them up just a little, and you'll need to be within a half-mile of any station trying to call you or any mobile trying to answer you.

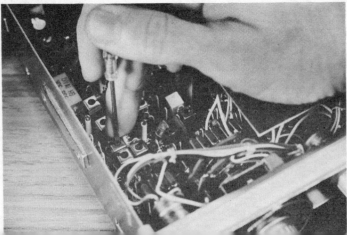

There's nothing wrong with buying your own crystals and installing them. But there are pitfalls.

First, be sure you know what kind (and frequency) of crystals to buy. They must match your transceiver more than just physically. Buy them from whoever makes or sells your brand of equipment.

Second, to operate legally, you must have the transmitter frequency checked whenever you or anyone else installs a new crystal.

Third, a new crystal may necessitate readjustment of certain coils that keep the transmitter functioning properly.

Fourth, don't try replacing crystals in a transceiver that uses a synthesizer (the 23-channel type). That's strictly a technician's job, requiring instruments for correct measurement and adjustment.

Where you can cause real trouble messing with coil adjustments is in the transmitter. In some units, the coils are sealed with beeswax. You can't turn one of them accidentally. In many transceivers, unfortunately, transmitter coils look just like receiver coils.

Transmitter misalignment can cause you more serious trouble than a receiver that's off-frequency. The transmitter might operate illegally, generating spurious emissions, interfering with other radio services, and creating communication difficulties for stations all around you. If you warp the transmitter off-frequency, it interferes with others and severely limits your own communications.

The worst of all is the likelihood of an FCC monitor picking up all this messy transmission. That will definitely get you an FCC notice of violation. You'll have to hire a licensed technician to repair and certify your transmitter; then you must send a report to the FCC that this has been done. If you've fouled up the transmitter with doodling, the repair will cost you more than if some simple defect had developed.

None of these cautions about keeping your fingers and screwdrivers out of your transceiver are meant to worry you. They're aimed at saving you money and helping you keep your transceivers always ready for their prime purpose: communications.

The best way around any of this is to find a good sales-and-service outlet for CB. A company that does both has a double interest in keeping your equipment operating well. That's what will draw you back when you're ready to buy newer and better equipment.

You may wonder why some CB repairs cost so much. Beyond the reasons already cited (the foul-ups an inexperienced doodler causes), there are a couple of other factors.

This benchful of expensive and complicated equipment illustrates one reason. No qualified technician will work on your transmitter without the proper instruments. You're looking at test apparatus that can accomplish six things: check *precisely* the frequency of every channel in your CB transmitter, measure the maximum modulation percentage of which your transmitter is capable, measure output power and input power of your transmitter, test the sensitivity of your receiver according to the 10 dB (S + N)/N standard, allow the technician to align every receiver and transmitter coil precisely for peak performance as well as legal operation, and certify (for the FCC if necessary) that a transmitter does not violate any technical regulation.

The transmitter technician working for a communications servicing business must be more than just qualified. He must also be licensed. The Federal Communications Commission requires that all measurements and adjustments that might affect technical operation of a transmitter be made only by an FCC-licensed technician. For CB and most commercial communications, a second- or first-class radiotelephone operator license is satisfactory. (Some communications equipment requires a radiotelegraph operator license.)

Before you let a technician work on your CB equipment, ask to see his framed license (it should be on the wall of his shop) or the license verification card he may carry in his wallet. That proves he's qualified, at least as far as the FCC is concerned.

With repairs properly certified, you can enjoy your Citizens Radio communicating with your mind at ease and with equipment in tip-top shape.

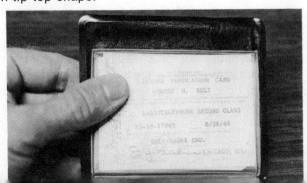